William Otis Crosby

Geological Collections

Dynamical geology and petrography

William Otis Crosby

Geological Collections
Dynamical geology and petrography

ISBN/EAN: 9783337218508

Printed in Europe, USA, Canada, Australia, Japan

Cover: Foto ©berggeist007 / pixelio.de

More available books at **www.hansebooks.com**

Guides to the Museum of the Boston Society of Natural History.

GEOLOGICAL COLLECTIONS.

DYNAMICAL GEOLOGY

AND

PETROGRAPHY.

By W. O. CROSBY,

ASSISTANT.

BOSTON:

PUBLISHED BY THE SOCIETY.

1892.

DEDICATION.

To the Memory of

DR. LUCY E. SEWALL,

A devosed and enthusiastic student of Geology,
whose friendship and example have been a constant iuspiration
to the author, this volume is gratefully inscribed.

PREFACE.

This volume is a Guide to the illustrations of Dynamical Geology in the Vestibule and of Petrography, including both Lithology and Petrology, in Room B. Visitors and students are advised to follow the order of the explanatory text, beginning with the Vestibule and completing the study of the geological agencies before passing to the petrographic collections in Room B. A part only of the main floor of this room is devoted to Petrography, sections 24 to 31 and the gallery being reserved, provisionally, for illustrations of the local and New England geology.

On entering Room B, the lithologic collection begins on the left hand, in section 1 of the wall-cases, and extends around the room to section 23 ; while the two central or floor-cases contain the petrologic collection, with the exception of some of the larger illustrations, which are in the table-cases and in the Vestibule.

The numbers on the doors of the cases indicate the sections, and, to facilitate reference to the text of the Guide, corresponding numbers have been placed at the head of each page. The numbers printed in small type in the upper left hand corner of the labels, on the specimen tablets, refer to the Society's general written catalogue, which, however, contains no information not already given on the labels or in the following pages. The numbers corresponding to those by which the specimens are designated in the text are printed in larger type on small rectangles of paper which are either attached directly to the tablet or to the label. These numbers are also made more conspicuous in the Guide by being printed in black-faced type.

The reference numbers do not, as in the mineralogical Guide, form one continuous series 'for the entire collection; but each section is numbered independently from 1 to 100, twenty numbers being assigned to each of the five shelves in a section, so that the specimens on the first or top shelf in any section always begin with No. 1, those on the second shelf with 21, on the third shelf with 41, the fourth shelf with 61 and the fifth or bottom shelf with 81.

Although a continuous series of reference numbers possesses certain obvious advantages, experience has shown that these do not compensate for the serious check which is thus placed upon the continued growth and improvement of the collections.

The large labels scattered through the collection correspond in a general way with the headings of the various divisions of the Guide. It will be observed that the system of labels and numbers is specially designed to make it easy to refer from specimens on the shelves to the explanation in the text, and *vice versa.* Throughout the Guide, the more detailed explanations, intended particularly for special students of mineralogy, have been printed in small type, a device which will facilitate the progress of the general visitor.

The publication of this Guide has been delayed for several years by the want of means; and it has been rendered possible at this time only through the generosity of Miss Marian Hovey.

ALPHEUS HYATT,

Curator.

TABLE OF CONTENTS.

INTRODUCTION.

Natural History includes the study of the earth and all its natural products, and also embraces Astronomy, in so far as this throws light upon the origin and history of the earth. The main object of the collections in this Museum is, therefore, to present as complete an epitome of the history of the earth and its present condition as the building and the slender means at the command of the Society will permit.

The broadest distinction afforded by terrestrial phenomena is that existing between inorganic and organic matter; and Natural History thus embraces two grand divisions: Geology and Biology. Each of these main branches includes three subdivisions — dynamical, structural, and historical, according as we regard the relations of the phenomena to force or energy, to space or to time.

Dynamical Geology treats of the forces now operating upon the inorganic materials of the earth — the geological causes or agencies — such as the currents and waves of the ocean, winds, rivers, rain, frost, earthquakes, volcanoes, hot springs, etc., and the various effects which they produce.

Structural Geology treats of the relations of the different kinds of minerals and rocks, and rock structures, produced by these agencies in the past; and, in its broadest aspect, takes account of the constitution and architecture of the entire crust of the earth.

Historical Geology treats of the relations of minerals and rocks in time, the historical succession of strata and their contained fossils (stratigraphy and paleontology), and also embraces all the changes or events of the earth's history.

Structural Geology divides naturally into two sciences : *Mineralogy*, which treats of the composition, structure, and physical properties of homogeneous chemical compounds or minerals ; and *Petrography*, which treats of the composition, structure, and distribution of rocks, or the massive and usually impure aggregates of minerals. Petrography is conveniently subdivided into the subordinate sciences of *Lithology* and *Petrology*, which are separately illustrated in the collections.

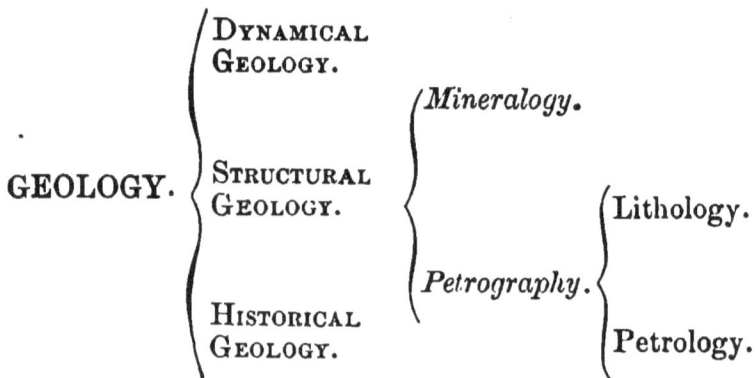

GEOLOGY.
- DYNAMICAL GEOLOGY.
- STRUCTURAL GEOLOGY.
  - *Mineralogy.*
  - *Petrography.*
    - Lithology.
    - Petrology.
- HISTORICAL GEOLOGY.

This volume, as stated on the title-page, is a guide to Dynamical Geology and Petrography, the Guide to the other division of Structural Geology — Mineralogy —

having been previously published as the first volume of the series; while the Guide to Historical Geology, including Stratigraphy and Paleontology, will form a third volume.

The illustrations of Dynamical Geology are in the cases on the right or west side of the Vestibule, and the illustrations of Dynamical Biology are similarly placed on the left or east side of the Vestibule. These are not, however, the only collections in the Vestibule. Besides the large slabs of sandstone showing fossil footprints and ripple-marks in the back part of the Vestibule, and the large mineral specimens, fossil tree, etc., on either side of the main stairs, which properly belong with the mineralogical and petrological collections in Rooms A and B, the window-spaces are occupied by a general or introductory collection. This collection is introductory to the dynamical collections and is designed (1) to call attention to those general features and relations of the earth a knowledge of which is essential to the comprehension of the nature and magnitude of the various terrestrial forces or agencies; and (2) to show the principal forms of both inorganic and organic matter upon which these forces operate. In other words, it embraces an outline or the rudiments of Physiography.

THE INTRODUCTORY COLLECTION.

ASTRONOMICAL RELATIONS OF THE EARTH.

This introductory collection properly begins with the more important astronomical relations of the earth, or with the earth as a planet, illustrating especially the relative distances, magnitudes, and motions of the sun and moon, which are the principal sources of the energy of the various agencies operating upon the earth's surface. The relative sizes and distances of these bodies are shown by the two models one of which is suspended in the upper part of each window case. The gilded ball represents the sun on the scale of about 225,000 miles to one inch, and the round black dot in the center of the square white plate shows the earth on the same scale, their diameters being approximately as 1 to 110. The distance between these models represents the mean distance of the earth and sun, about 93,000,000 miles, on' the same scale as the models; while the orbit of the moon is represented by a circle on the square plate, at its proper relative distance from the earth, about 240,000 miles. The moon itself, on this scale, would be microscopic, being represented by a dot having one fourth the diameter or one sixteenth the area of that representing the earth.

---

[1] These two sections are to be regarded as one, the specimens on each shelf continuing across from left to right; with one set of numbers, the odd decades being on the left and the even decades on the right.

The earth fills $\frac{1}{2,300,000000}$ σ part of the solar sky, *i. e.*, it intercepts that almost infinitesimal fraction of the light and heat which emanate from the sun on all sides. The corresponding fraction in the case of the moon is about $\frac{1}{1500}$. The sun and moon are interesting to us chiefly, not by virtue of their vast dimensions and distances, but as sources of force or energy; and we can form some conception of their stupendous importance in this regard by comparing the fractions given above with the energy actually intercepted by the earth. This energy comes to us chiefly in two forms : *gravitation* and *heat.*

It is gravitation that holds the earth and moon in their orbits, the mutual attraction of the sun and earth for example, being equivalent to a powerful cord connecting the two bodies and preventing the earth from moving off in a straight line. The strength of this invisible restraining force is as great as if every square foot of the earth's surface lighted by the sun's rays were connected with the sun by a steel rod between one third and one half of an inch in diameter. The attraction of the sun and moon also produces the oceanic tides.

The amount of solar heat received annually by the earth is sufficient to melt a layer of ice 136½ feet thick, covering the entire surface of the earth. Or, expressed mechanically, it is equivalent to more than two horse-power acting continuously on each square meter of the earth's surface. On the surface of the sun itself, the solar radiation is equal to the continuous action of more than 100,000 horse power per square meter.

The model over the right or west window-case illustrates chiefly the relative positions and motions of the

sun, earth, and moon. The large central globe represents the sun; and on turning the crank the smaller
globe, representing the earth, is carried around it; while
the smallest ball, representing the moon, is carried
around its primary, the earth. The mechanism is so
adjusted that the distance between the earth and sun
varies with the seasons, being greatest at the summer
solstice (June 21) and least at the winter solstice (Dec.
21). The model of the earth is suspended in such a
manner that its axis remains parallel to itself, *i. e.*,
maintains a constant direction in all phases of its annual
revolution around the sun; so that, the earth's equator
being inclined at an angle of $23\frac{1}{2}°$ to the ecliptic or the
plane of the earth's orbit, the north pole is constantly
turned toward the sun in our summer (Mar. — Sept.)
and the south pole in our winter (Sept. — Mar.)
The apparatus is always adjusted for the current month,
and thus illustrates and explains the changes of the
seasons.

The earth is not a perfect sphere but is flattened at the poles
the polar diameter or the earth's axis (7,899 miles), being about
$26\frac{1}{2}$ miles shorter than the equatorial diameter (7,926 miles).
The earth is, in other words, an oblate spheroid; but, although
its variation from the true spherical form seems actually large,
the flattening at the poles is relatively so small that an exact
model of the earth could not be easily distinguished from a perfect sphere.

INTERNAL CONSTITUTION OF THE EARTH.

Before noticing the surface configuration of the earth,
we may observe the most probable constitution of its
interior.

The weight of the earth has been determined by entirely independent experiments with the plumb-line, the pendulum and the torsion balance; and the closely accordant results which they afford show that the earth is about five and a half times heavier than a globe of water of the same size; $i, e.$, the mean density or specific gravity of the earth is about 5.5. But the average density of the materials composing the superficial portions of the earth, the common minerals and rocks, is less than half as great, or 2.6. Hence it is evident that the interior of the earth must consist of much heavier materials than the exterior.

There are several distinct lines of evidence pointing to the conclusion now generally accepted that the earth's interior is mainly composed of one substance — iron. Many of the volcanic rocks, and especially those which, from their homogeneity, we may suppose to have come from the greatest depth in the earth, contain a larger proportion of iron than is found in the materials normally forming the earth's surface. This is shown by the first series of specimens in this case (**12-16**). Gneiss (**12**) is taken as the typical representative of the earth's crust or surface, with less than five per cent. of iron. Then comes diabase (**13**) as the most representative eruptive rock, with from 5 to 15 per cent. of iron. And, lastly, we have the so-called ultrabasic eruptive rocks represented by the porphyritic magnetite from Cumberland, R. I., (**14**), containing 15 to 40 or more per cent. of metallic iron, or so large a proportion that they are essentially ores of iron. The meteorites (**15-16**), which, as the specimens show, are either pure iron or highly ferruginous, are believed to be, in a general way, average samples of the materials of which the earth is composed.

The density of the earth as a whole is readily explained, as Professor Dana has shown, if we assume that it is two thirds iron, or iron from the center to within 500 miles of the surface. And in the present state of the science we are justified in thinking of the earth as possibly a ball of iron some 7000 miles in diameter, surrounded by a thick layer of stony material, in which the percentage of iron and density diminish gradually toward the surface.

It seems inexpedient, in the present state of our knowledge, to attempt to illustrate the probable distribution of heat and of solid and liquid materials in the interior of the earth. Any views on these subjects are necessarily largely speculative; and the following statements are designed merely to show the narrow limits of our actual knowledge, and the general tendency of geological opinions on these subjects at the present time.

Numerous observations in mines, artesian wells, etc., show that the temperature of the ground always increases downwards from the surface; and the much higher temperatures of hot springs and volcanoes show that the heat continues to increase to a great depth, and is not merely a superficial phenomenon. The observed rate of increase is not uniform, but it seldom varies far from the average, which is about 1° Fahr. per 53 feet of vertical descent, or, in round numbers, 100° per mile. This rate, if continued, would give a very high temperature at points only a few miles below the surface; and until within a few years, the idea was generally accepted by geologists that the increase of temperature is sensibly uniform for an indefinite distance downward; that in the central regions of the earth the temperature is far higher than anything we can conceive, and that everywhere below a depth of 20 to 40 miles the temperature is above the fusing point of all rocks; and hence that the earth is an incandescent liquid globe covered by a thin shell or crust of cold, solid rock.

There are, however, several important considerations which now oblige us to modify this view. It is certain that the temperature of the earth cannot increase downward at a uniform rate, but only at a constantly and rapidly diminishing rate; and

It is probable that below a depth of 300 miles the temperature is everywhere sensibly the same, and nowhere incomparably higher than temperatures we are acquainted with on the earth's surface. The lavas emitted by volcanoes in all parts of the globe prove that the subterranean temperature is sufficient to fuse ordinary rocks; but many geologists hold that the enormous pressure to which matter must be subjected in the deeper portions of the earth, and which is supposed to favor solidification, neutralizes the high temperature and keeps the great central mass of the earth solid in spite of the heat. The existence of the oceanic tides and certain astronomical phenomena are believed to support this view. All that we actually know, however, is that the temperature of the ground increases downward until it becomes sufficient to fuse considerable volumes of rock.

EXTERNAL CONFIGURATION OF THE EARTH.

The inequality of the earth's polar and equatorial diameters is relatively small ; but the inequalities of relief presented by its surface, and which we call continents and oceans, mountains and valleys, etc., are utterly insignificant compared with the vast dimensions of the earth itself. This is clearly illustrated by the two large diagrams on the wall in the west window-space. Each diagram represents a sector of the earth one degree or about 70 miles broad. The first one is drawn on a radius of 8 inches or a scale of 500 miles to the inch ; the middle circle represents the surface of the earth, while the inner circle marks the thickness of the earth's crust, and the outer circle the upper limit of the atmosphere. The colored portion of this diagram is all that is included in the second diagram, which is drawn on a radius of 800 inches or a scale of 5 miles to the inch, and shows the thickness of the solid crust of the earth (assumed as 35 miles), the hight of the atmosphere (also assumed as 35 miles), the probable extreme depth of the ocean (about 30,000 feet), the probable mean depth of the ocean (about 15,000 feet), the mean hight of the land (about 1,000 feet) and the extreme hight of the land (29,000 feet). To fully appreciate the diagram it is necessary to remember that the two converging vertical lines representing the radii of the earth would meet at a distance of 800 inches or $66\frac{2}{3}$ feet from the blue line which represents the surface of the ocean.

The area of the earth's surface is 197,000,000 square miles, of which 144,500,000 are water and 52,500,000 are land, the proportion being nearly 275 to 100, or approximately 8 parts water to 3 parts land. The mean depth of the ocean greatly exceeds the mean hight of the land, so that while the land has nearly three eights the area, it has only one fortieth the volume of the ocean. The general distribution of land and water is shown by the physiographic map of the world on this wall-space.

The general contours or vertical forms of the continents and ocean-basins are also shown on this map by the shading, the boundaries of the different tints being, as explained on the map itself, essentially contour lines. This map shows further that the actual shore-line is not necessarily the true border of the continent and ocean. Along some coasts, as that of New Jersey, the true continental edge is submerged, and the water deepens very slowly at first, so that the 100 fathom line is 50 to 150 miles from the shore, while a few miles farther brings us to the 500 fathom line ; the more abrupt descent indica-

ing that we have crossed the true boundary line between
the continent and ocean basin.

In accordance with this principle, all islands standing on
these submerged borders of the continents are classed as continental, while only those rising from the deep waters of the
ocean are properly oceanic. Thus Great Britain and Ireland are
a part of the European continent; Newfoundland and the Grand
Banks belong to North America, and the Malay Archipelago to
Asia.

The positions and directions of the principal mountain-
ranges of the globe are indicated by the heavy black lines
on the small maps of the continents; and it will be observed that these axes of elevation are, for the most
part, grouped near to and parallel with the coast-lines.
These small maps and the accompanying sections of the
continents also illustrate the following general laws in
the reliefs or surface forms of the continents: (1) The
continents have in general elevated mountain borders
and a low or basin-like interior. (2) The higher border
faces the larger ocean.

The distribution of land in high latitudes, so far as
known, is shown more clearly by the maps of the polar
regions in the case (6); and the physiography of the
United States is represented in greater detail by the
colored map at the top of the case (2) and the shaded
map below it (3). The latter represents a model of
the United States viewed from above, and also the
middle belt as seen obliquely from the south, bringing
out the east-west profile more distinctly. Far superior
to these illustrations, however, is the model itself — the

large relief map of the United States. This is constructed so as to show the natural curvature of the
earth's surface; and, except that the vertical scale is
necessarily exaggerated, it is a faithful representation
of the main relief features of our country. The longest
diameter of the model, it will be seen, embraces but a
small section of a complete circle; and it thus shows
almost as clearly as a globe how small a part of the terrestrial sphere the United States actually covers. The
model, which is in a large degree self-explanatory and
will repay careful study, shows not only the configuration of the land, but also the subaqueous contours of the
great lakes and the adjacent parts of the ocean, thus
developing the true form and relief of this part of the
continent, and showing, as previously pointed out, that
the continental border is often outside of the actual
shore line. The mean annual isothermal lines are also
traced upon the map.

CIRCULATION OF THE ATMOSPHERE — ATMOS-PHERIC CURRENTS OR WINDS.

The wind map (4) illustrates the circulation of the
atmosphere, or the prevailing winds of the globe. The
atmosphere, under the influence of gravity, tends always to adjust itself so as to be in a state of equilibrium,
the chief condition of which is a uniform density at any
given altitude, the density diminishing upward with
the decreasing pressure. But the continuance of this
equilibrium requires uniformity of temperature, as well
as of pressure, in all parts of the same stratum of air.

If any stratum is unequally heated in different parts, the equilibrium is destroyed. The warmer portion expands and becomes lighter; and being pressed upon by the adjacent colder and heavier air, it rises, and its place is occupied by the latter. This process results in an ascending current from the region of greatest heat, and horizontal currents flowing from all directions towards that region.

Hence from the permanent inequality in the distribution of heat in the tropical and polar regions, there results, in each hemisphere, a constant circulation of the atmosphere consisting of (1) an *ascending current*, in the zone of highest average temperature; (2) a *polar current* flowing upon the surface, from each pole towards the equator; and (3) a *return current* flowing from the equator towards either pole partly in the upper air and partly on the surface.

The upward motion of the ascending current is imperceptible, the atmosphere seeming to be in a state of rest; hence this is known as the zone of equatorial calms. Near the tropics, where the return currents descend, there are also belts of calms; but these are less defined than the equatorial belt. The three belts of calms are represented by the yellow color on the map.

In consequence of the earth's rotation, the normal direction of the polar current is northeasterly in the northern hemisphere, and southeasterly in the southern; and the return currents become southwesterly winds in the northern hemisphere and northwesterly in the southern. These directions are plainly indicated by the arrows on the map, and also by those within the circumference of the circle in the diagram.

Since the earth performs one entire revolution on its axis every twenty-four hours, the velocity of rotation at the equator must be somewhat more than 1,000 miles an hour. As each successive parallel towards the poles has a less circumference than the preceding, the velocity of rotation diminishes with increasing latitude, until at the poles it is zero. If, therefore, particles of air move from the polar regions towards the equator, each step in advance will bring them upon parallels where the rotation is more and more rapid. The new velocity cannot be instantly acquired; consequently, at each successive parallel the moving particles are left a little behind, or to the west of their previous position; and when they reach the tropics they are many degrees west of the meridians upon which they left the polar regions. A similar cause operates to give the return currents their eastward tendency. The particles moving towards the poles find, at each successive parallel, a rotary velocity less than at the preceding. Not acquiring the new and less rapid motion instantaneously they gain a little at each parallel, and find themselves in advance or to the east of their former position.

The general law of atmospheric circulation explained above gives rise to three distinctly marked wind zones on each side of the equator; namely: — (1) the zone of *constant* or *trade-winds*, extending to latitude 25° or 30°, represented by the pink color on the oceanic portions of the map; (2) the zone of *variable winds*, with alternate polar and equatorial currents dominating, extending to latitude 60°, or near the arctic circle, and indicated by the green color on the map; and (3) the zone of prevailing, though not constant, *polar winds.*

The positions of the various wind and calm zones change with the seasons, all advancing northward and retiring southward with the apparent motion of the sun. The northern and southern limits of the trade-winds and equatorial calm belt, for the four seasons of the year, are shown in the diagram in the lower left-hand corner of the map.

CIRCULATION OF THE OCEAN-MARINE CURRENTS.

The next map (7) illustrates the oceanic circulation, the principal currents being represented by the bands of light and dark green, and their directions by the arrows.

The main causes of these vast movements in the ocean are found in the winds, the excessive evaporation within the tropics which tends to lower the level of the water there, and the differing temperatures of the polar and equatorial regions.

Two series of currents of opposite character pervade the sea in high latitudes : the *cold*, flowing from the polar regions toward the equator, and the *warm*, flowing in the opposite direction, the tendency of both currents being to restore the equilibrium distributed by the trade-winds, the intense heat and the evaporation within the tropics. In the middle latitudes, where the opposing currents meet, the cold, being heavier, sink beneath the warm and disappear, continuing their course in the deep waters. These cold under-currents, having reached the inter-tropical seas, gradually rise again to the surface, where they become heated ; and contributing their waters to the great equatorial current, which flows westward on each side of the equator, they finally return to the poles.

The westward flowing equatorial currents are due chiefly to the incessant action of the trade-winds ; and, if they were not intercepted by the continents, they would continue in the same direction quite around the globe. They are, however, deflected by the eastern shores of the continents and thus give rise to the well-defined return currents. Hence the general or normal currents of the ocean are of three classes : the equatorial currents, owing their impulse and direction to the trade-winds; the return currents, resulting from the deflection of these, and, in consequence of the earth's rotation, moving northeasterly in the northern hemisphere and southeasterly in the southern ; and the polar currents, which, like the polar atmospheric currents, flow normally in directions opposite to those of the return currents.[1]

The polar and return currents may exist in all longitudes, but are appreciable only toward the sides of the oceans. A portion of the return current in each ocean continues in its northeasterly trend to the polar regions; but the main part takes a more easterly course in middle latitudes, and, becoming cooled, returns along the eastern border of the ocean towards the equator. The equatorial and return currents together thus constitute a vast elliptical movement of the waters of the ocean,— westward in the torrid and eastward in the temperate zone. In the center of each ellipse is a broad expanse of quiet water, which is covered to a great extent with seaweed, forming the so-called Sargasso seas.

In the Atlantic and Pacific oceans the equatorial currents are really double, including a north equatorial generated by the northeast trade-wind, and the south equatorial owing its impulse to the southeast trade-wind, the two currents being separated by the equatorial belt of calms. Where these equatorial

[1] An explanation of the oblique directions of the polar and return currents is given in the account of the atmospheric circulation on page 14.

currents impinge upon the eastern shores of the continents and
are deflected to form the return currents, we find the swiftest
and most sharply defined of all these rivers of the ocean. The
Gulf Stream, which sweeps the eastern coast of North America,
is an exceptionally striking example. The map shows that in
the Atlantic Ocean the north equatorial and a very large part of
the south equatorial strike the coast of South America north of
Cape St. Roque and are deflected northward, flowing through
the Caribbean Sea into the Gulf of Mexico, whence they issue
through the Straits of Florida as the Gulf Stream, with a temperature of 80° Fahr., a velocity of 5 miles per hour, a breadth
of 32 miles and a depth of more than 2000 feet. This means
that more than five trillions of cubic feet of warm water are
discharged from the Gulf of Mexico into the North Atlantic
every hour. As it advances northward, it becomes broader and
shallower; and in the latitude of Cape Hatteras it begins to bear
away from the coast and crosses to the Azore Islands. The
corresponding current in the North Pacific is known as the Kuro
Sivo or Japanese current.[1]

This map also shows the main divisions of the continental drainage, the colors indicating the land areas
tributary to each of the oceans. The predominance in
this respect of the Atlantic basin over the much larger
basin of the Pacific is especially noteworthy.

DISTRIBUTION OF TEMPERATURE ON THE EARTH'S SURFACE.

The temperature map (1) shows, by means of annual
isothermal lines, or lines of equal annual temperature, the
general distribution of temperature or solar heat upon
the earth. The frigid, temperate, and torrid zones of

[1] For an explanation and illustrations of tidal phenomena — tidal waves
and currents — students are referred to the standard works on physical
geography and physiography.

temperature are also shown by bands of different colors. All points located on the same isothermal line have approximately the same mean annual temperature; and all portions of the same color area may be regarded as belonging to the same climatic zone of the earth. The zones of temperature, however, coincide only very imperfectly with the geographic zones.

These deflections of the isotherms are due partly to the elevations of the land, the mountains and plateaus being lifted above the lower and denser portion of the atmosphere which is most efficient in absorbing and retaining the solar heat; and the influence of the ocean currents upon the distribution of temperature is seen very clearly by comparing this map with the map of the oceanic circulation. The comparison shows that where the currents of cold water from high latitudes flow toward the equator, as on the east coasts of North America and Asia, the isotherms are crowded together and the temperate zone becomes very narrow, so that a few degrees in latitude corresponds to a marked change in climate. While the warm currents flowing from low to high latitudes have just the contrary effect. This is particularly evident in the case of the Gulf Stream, which flows from the Gulf of Mexico northeasterly through the North Atlantic Ocean; expanding the temperate zone, and giving Iceland and Scandinavia a climate as mild as that of Newfoundland. In other words, the map shows that, as a result of oceanic movements, the temperate zone is narrow on the west side of each ocean and broad on the east side, while the torrid zone of temperature is broad on the west side and narrow on the east.

It is estimated that the heat carried by the Gulf Stream alone from the torrid into the temperate and frigid zones is equal to one fourth of all that received directly from the sun by the entire Arctic regions, or as much as is received from the sun by 1,560,000 square miles at the equator.

The circulation of the atmosphere, as shown on the wind map, is in accordance with the same general plan as that of the ocean, the ocean currents and prevailing winds being generated and controlled by the same great causes,—the solar heat and the earth's rotation; and the atmospheric circulation must tend to modify the distribution of the solar heat in essentially the same way as the oceanic circulation, but not to the same extent. It is, indeed, probable that a greater quantity of heat is conveyed by the Gulf Stream alone from the tropical to the temperate and Arctic regions than by all the aerial currents which flow from the equator.

The two small maps on this map show the isotherms for July and January, respectively the hottest and coldest months of the year. Together with the figures accompanying the names of places on the main map, they show the annual range of temperature of different localities, due partly to the astronomical change of the seasons, but very largely to the fact that the land absorbs the solar heat rapidly in summer and the surface soon attains a high temperature, while in winter it is as rapidly los by radiation and a comparatively low temperature is soon reached. Thus the isotherm of 70^0 moves from the West Indies in winter to the St. Lawrence River and the Great Lakes in summer.

The waters of the ocean, on the other hand, on account of their capacity for heat, and constant circulation, are subject to a comparatively slight annual range of temperature. Hence the oceanic or insular climate is much more equable than the continental, and the annual north-south swing of the isotherms, as the maps show, is very much more restricted on the ocean than on the land. On the coasts we find a mingling of continental and oceanic climates. Thus at St. Paul, in the heart of the continent, the average temperatures for January and July are 14^0 and 73^0; at Boston they are 26^0 and 71^0; and at the Azore Islands they are 59^0 and 68^0.[1]

[1] For illustrations of the surface distribution of the subterranean or volcanic heat and of earthquake phenomena the student is referred to the dynamical collection, which follows n the next case.

DISTRIBUTION OF RAIN AND SNOW.

The rain map (**5**) illustrates the main facts in the distribution of the annual rainfall or precipitation of moisture over the globe. The amount of the average annual precipitation is indicated in a general way by the depth of the blue color, and more exactly for particular localities by the figures near the names giving the amounts in inches.

Water, whether in the sea or on land, is slowly transformed into invisible vapor, which, being much lighter than air — as 3 to 5, — rises, and is diffused through every part of the atmosphere. Thus the latter becomes the great reservoir of aqueous vapor. The capacity of the air for the absorption of vapor increases with its temperature; but, at any given temperature, there is a certain limit beyond which it can receive no more. When filled to its utmost capacity it is said to be *saturated* with humidity, and the least lowering of the temperature causes a condensation of moisture in the form of dew, fog, clouds, or rain; but if the temperature is raised, the capacity for vapor being increased, absorption recommences.

Condensation and rain are mostly caused by the cooling of currents of warm air laden with aqueous vapor. A *warm wind* setting from the tropics clear and dry, and advancing into cooler latitudes, continually diminishes in temperature. Hence, without receiving additional vapor, its relative humidity constantly increases until saturation is reached, when the air becomes moist and cloudy, and finally rain falls.

A *cold wind*, on the contrary, starting from the polar regions saturated with vapor and full of clouds, and advancing to warmer latitudes, has its capacity for moisture constantly increased. Hence it becomes at every step less humid, and its clouds dissolve, leaving the air clear and dry.

Thus warm winds, blowing towards cooler regions, bring rain; while cold winds, advancing to warm climates, bring fair weather and drought.

Ascending currents of air give rise to similar phenomena. The warm air, laden with vapor, rising into the upper regions of the atmosphere, expands and becomes cool; and its vapor, condensing rapidly, returns to the ground in copious showers. Thus are produced the rains of intertropical regions, and the summer showers of middle latitudes. The arrows on the map give the directions of the principal winds bringing rain.

Mountain-chains, in general, act as condensers, especially when lying across the path of warm winds. Upon the side exposed to the wind, the air is forced upward along the slopes, and its vapors are condensed into clouds, whence torrents of rain fall; while on the opposite side it descends, with increasing temperature, as a dry wind.

Thus the eastern slope of the Andes intercepts the vapors borne by the southeast trade-winds of South America and thus divide the moist plains and luxuriant forests of the Amazon and Paraguay basins, from the rainless and barren coast of Peru and

Chili.   The Sierra Nevada and Cascade Ranges along the
northwest coast of North America intercept, in a similar manner, the warm and moist return trades of the Pacific, while the
interior plains of the continent are dry and barren.   In Europe,
the mountains of Scandinavia and the British Islands condense
the principal part of the vapor brought by the winds from the
Gulf Stream, and the countries lying to the eastward are, in
consequence, relatively dry.

It follows from this principle that the borders of the
continents, especially when they face warm winds from
the oceans are, in general, more moist and fertile than
the inland districts.   The Mississippi valley is an exception, because no bordering mountains separate it from
the return trades, which are laden with vapor during
their passage across the Gulf of Mexico.   The interior
of South America is also largely an exception, since, for
the most part, there are no prominent mountains along
the eastern coast.

Although the distribution of rain is much more liable than the
temperature to irregularities and extremes dependent upon local
circumstances, yet the following general laws hold good : — 1.
The *average quantity of rain falling annually* is greatest in the
tropical regions, where the rapidity of evaporation, and the
absolute amount of vapor in the air, are both extreme ; and the
average decreases gradually with increasing latitude, as shown
by the following table, and by the diagram in the lower right
hand corner of the map.

Latitude.	Inches of Rain.	Latitude.	Inches of Rain.
0° (Equator)	100	50°	30
20°	80	60°	20
30°	60	70°	10
40°	40	80°	5

2. The average rainfall diminishes gradually from the coasts to the interior of the continents, as shown by the map and also by the following table for the old world: —

Countries.	Average Rainfall.	Rainy Days.	Countries.	Average Rainfall.	Rainy Days.
British Isles.	32 inches	156	Hungary.	17 inches	111
Western France	25 "	152	Russia(Kas'n	14 "	90
Eastern France.	22 "	147	Siberia		
Germany.	20 "	150	(Yakutsk.)	10 "	60

Owing to the intimate relation which exists between the temperature, the winds, and the condensation of vapor, each of the great climatic zones of temperature and winds has also a distinct system of rains. These rain-zones are bounded by the red and blue lines on the map, and the inscriptions sufficiently explain the characteristics of each.

The broken blue lines on the map show the equatorial limits of the fall of snow at the level of the sea. That is, between these lines and the equator snow never falls at the level of the sea, but only on the high lands or mountains.

COMPOSITION OF THE EARTH.

This introductory collection would be incomplete without the general illustration of the inorganic materials composing the earth afforded by the remaining specimens in the west window case, including some of the more important kinds of rocks and their component minerals.

Minerals are the homogeneous, chemical substances making up the whole of the inorganic portion of the earth. The earth is not, however, a uniform mixture of all the various kinds of minerals of which it is composed, but these are grouped or associated in different ways in different parts of the earth's crust; and these great masses or layers of the crust differing in mineral composition are the rocks. In other words, rocks are more or less definite aggregates of minerals; being made up of crystals or grains of one mineral only, like limestone, or, more usually, of several minerals, like granite.

This is not the proper place to notice in detail the composition and characteristics of even the few kinds of minerals and rocks selected for this illustration. The most important distinction observed in a general view of the composition of the earth is that existing between what are conveniently called its original and its secondary constituents. By the original constituents we mean, broadly speaking, those composing the primitive crust of the earth, resulting from the original solidification of the incandescent globe, or formed during all subsequent ages by the solidification of the fused materials or lavas which have flowed out from the interior through fissures in the crust. These are the igneous constituents, or those in the formation of which heat has been the chief agent, and they are mainly anhydrous, basic silicates, such as the feldspars, micas, hornblende, augite, etc.; although including some oxides, like magnetite and quartz, some sulphides, like pyrite, at least one phosphate, apatite, and one very important native element. iron; but no carbonates, sulphates, or chlorides, and no hydrous compounds.

These original or igneous rocks and minerals are the materials upon which the geological agencies explained in the preceding pages operated in the earliest period of the earth's history, and, in a diminishing degree, during

all subsequent periods down to the present. And they
are the source of all the secondary minerals and rocks.
Although very stable compounds in the presence of heat,
the most of the original or igneous rocks are readily decomposed and disintegrated by the chemical and mechanical action of water, frost, and the various constituents of
the atmosphere. The hard, anhydrous silicates like the
feldspars, hornblende, mica, etc. are thus changed to
softer, hydrous silicates like kaolinite, chlorite, etc., to
quartz, and to carbonates, sulphates, chlorides, and
hydrous oxides. These secondary compounds and fragments, when distributed by the sorting action of water
and organic matter, and subjected more or less to the
action of heat and pressure in the earth's crust, have
yielded the various kinds of secondary or aqueous rocks,
such as sandstone, slate, limestone, gneiss, etc. The
earth's surface, in the earliest geological times, must
have been made up wholly of the original minerals and
rocks, but it is now chiefly composed of the secondary
kinds ; and these have been derived in large part from
older secondary rocks, the same material being worked
over and over again by the aqueous agents. We may,
therefore, in conclusion, conceive the earth's crust as
consisting of a vast thickness of the still unaltered, original or igneous rocks, covered by a lesser but great thickness of the secondary or aqueous rocks in more or less
regular layers or strata, with numerous masses and layers
of igneous rocks which have been at different times
forced up through and among the aqueous strata.

The specimens in the lower half of this case illustrate
the preceding paragraphs. First come, on the second

main shelf, the mechanically formed sedimentary rocks
(**21–29**) and the chemically and organically formed sedimentary rocks (**31--39**) ; followed, on the third shelf, by
the metamorphic or crystalline stratified rocks, including
the schists (**41 -47**) and the gneisses (**51-57**). From the
gneisses the passage is natural to the plutonic division of
the original or eruptive rocks on the fourth shelf, the
acid types being placed on the left (**61–64**), and the
basic on the right (**71–74**) ; and on the bottom shelf we
come to the true volcanic rocks or lavas, which are arranged in the same way, the acid species on the left
(**81–87**) and the basic species on the right (**91–97**).

DYNAMICAL GEOLOGY.

Dynamical Geology is that branch of the science which treats of the processes of geological change at present in progress; and it is essentially the foundation of geology, since we must work continually from the known to the unknown, finding in the causes of changes now taking place a key for the interpretation of those changes which have occurred in the past.

Unfortunately, however, it is also true that no other department of geology is so difficult to treat satisfactorily in a museum. The agencies themselves, in operation, cannot be shown; and they can be only very inadequately represented pictorially or by models. Hence our only resource in most cases is to illustrate the geological causes, so far as practicable, by examples of their recent effects, and leave the remainder to the imagination, aided by the descriptions of the standard authors, and, still better, by the visitor's recollections of personal observation on the seashore and mountains, or in the neighborhood of rivers, volcanoes, and glaciers.

The geological agencies are divided naturally into two great classes: (1) those operating below the earth's surface, the subterranean or igneous agencies, deriving their energy from the central heat of the earth; and (2) those operating upon the earth's surface, the superficial or aqueous agencies, deriving their energy from the solar heat.

In the earlier periods of geological time, when the
outer portion of the earth had a much higher temperature than at present, so that the water of the globe could
exist only in the form of vapor in the atmosphere, the
chemical and mechanical changes taking place, even on
the earth's surface, must have originated chiefly in the
subterranean heat.   But as the outer layers of the globe
cooled, and the changes due to the internal heat gradually subsided, the influence of the solar heat must have
become more marked, giving rise to that wide circle of
superficial changes in which the circulation of air and
water over the earth, due to variations of temperature,
is such an important factor.   It is thus apparent that we
simply follow the natural or historical sequence in considering the subterranean before the superficial agencies.

The dynamical collection is in the two cases on the
west side of the vestibule (right hand on entering), the
illustrations of the subterranean agencies occupying the
right hand case and the illustrations of the superficial
agencies the left hand case.

SUBTERRANEAN OR IGNEOUS AGENCIES.

HORIZONTAL COMPRESSION AND CORRUGATION OF THE EARTH'S CRUST.

The principal reasons for believing that the temperature of the ground increases downwards, and that the
earth's interior is very hot, a great reservoir of heat,
have been stated in the introduction.   The earth is not

only a very hot body, but it is rotating through intensely cold space, and, therefore, must be a cooling body, the interior heat slowly passing out toward the surface and being radiated into space. Cooling means contraction, and consequently the heated interior of the earth, as it cools, must constantly tend to shrink away from the cold external crust. Of course no actual separation of the crust and interior can take place, but there is no doubt that the crust is left unsupported to a certain extent, and it must then behave like an arch with a radius of 4,000 miles, and the result is an enormous horizontal or tangential pressure.

It is now believed, however, that an entirely independent cause also tends to develop a powerful tangential strain in the earth's crust. It is well known that the centrifugal force arising from the earth's rotation is sufficient to change the otherwise spherical form of the earth to an oblate spheriod, with a difference of about twenty-six miles between the polar and equatorial diameters. It is also well known that while the earth turns from west to east on its axis, the tidal wave due to the attraction of the moon moves around the globe from east to west, thus acting like a powerful friction brake to stop the earth's rotation. Our day is consequently lengthening, and the earth's form as gradually approaching the perfect sphere. This means a very decided shortening and consequent crumpling of the equatorial circumference, and is equivalent to a marked shrinkage of the earth's interior, so far as the equatorial regions are concerned.

The subterranean agencies are chiefly various manifestations of this enormous horizontal pressure, which is one of the most important and generally accepted facts in

geology. The principal and most direct result of the
lateral thrust, whether due to cooling or tidal friction, is
the corrugation or wrinkling of the crust; and the earth-
wrinkles are, generally speaking, of three orders of magnitude,— continents, mountain-ranges, and rock-folds or
arches.

That lateral or edgewise pressure does produce undulations or wrinkles, alternate elevations and depressions, in layer of flexible material is a matter of common
experience, as with the folds and wrinkles of our cloth-
ing. This is shown by the thick layers of felt in the
model (25). The layers were first spread out horizontally and then squeezed endwise by shoving the confining
blocks toward each other. The next model (24) is a
farther illustration of the same principle, showing in
plaster the result of an experiment by Alphonse Favre
made by spreading layers of plastic clay on a stretched
sheet of India rubber and then allowing the rubber to
contract, carrying the clay with it and forcing the layers
into irregular folds or corrugations. In this experiment
the contraction of the rubber represented the shrinkage
of the earth's interior due to cooling.

Many features of the rocks composing the earth's crust
indicate more or less plainly that they have been exposed
to an enormous horizontal pressure. The most of these
will be fully illustrated and explained in the petrographic
collections in Room B; and it will suffice here to show
only the plainest proofs of this tangential compression.

The fossil shell (22) was originally circular, but it has
been compressed to a very flat form. The specimens on
the next tablet (21) are pebbles from the Roxbury con-

glomerate whose forms have been distorted and indented
by mutual compression, showing that this rock formation has experienced a powerful lateral squeezing. And
the specimen of contorted shale (23) shows that when
the rocks are composed of flexible layers they are wrinkled and plicated in the same manner as the layers of
clay and felt in the artificial examples. These general
illustrations, which are found in nearly all parts of the
earth's crust, show that the lateral thrust required by the
theories of internal cooling and tidal friction is an established fact.

ELEVATION AND SUBSIDENCE OF THE EARTH'S CRUST.

Relative changes of level of the land and sea are the
most important of all subterranean phenomena, the influence which these vertical movements of the crust
exert upon the forms, sizes and disposition of the continents and ocean-basins, and the general configuration
of the earth's surface, far exceeding the combined effect
of volcanoes and earthquakes; although these latter phenomena are far more obvious to the senses and more
familiar to the mass of mankind.

Geologists now generally believe that the surface of the ocean
is essentially stationary, and that any relative changes of level
of land and sea are due to the movements of the former. The
exact nature and cause of the movements are unknown in most
cases; but it is probable that the lateral pressure in the earth's
crust usually acts directly or indirectly. According to this
view, continents and ocean-basins are great upward and down-

ward bendings or undulations of the earth's crust resulting from the attempt of the crust to adapt itself to the shrunken interior. There are good reasons for believing, however, that some of the less important movements of the crust, especially in volcanic regions, are caused by expansion or contraction resulting from a gain or loss of heat in the part of the crust affected.

Slight movements of elevation or depression are sometimes accomplished suddenly and are accompanied by earthquakes; but all important oscillations of the crust take place with extreme slowness, a few inches or a few feet in a decade or in a century. Small changes of level are not easily detected except along the seashore, and even here the evidence is much more conclusive for elevation than for subsidence; since the land, in sinking beneath the sea, carries with it the evidence that it was once above the sea.

The principal proofs of vertical movements of the land are : —

(1) *Changes in position of works of man with respect to the sea-level.* The ruins of the Roman temple of Serapis at Pozzuoli, near Naples, shown in the picture (**1**), are a classic and perfect example of this kind of evidence. Only the floor and three columns of this structure still remain in their original positions; and when first discovered the floor and the lower part of the columns were covered with the debris of the temple and marine sediment. Above the part thus protected, the columns, as indicated in the picture, were perforated, to a hight of twenty feet, by numerous borings made by the bivalve mollusks called *Litho-domus* (lithos, a stone; and domus, a house), because they bore holes in the rocks near the water-line. The

floor of the temple was, of course, originally above the level of the sea. Then a subsidence, so gradual as to be unnoticed by the inhabitants, took place. Italian historians state that in 1530 the sea covered the site of the temple; and the lithodomus-borings on the columns show the extent of the subsidence, the floor of the temple being carried twenty feet below the level of the sea. The land then began to rise until, at the time the ruins were discovered, the base of the temple was several feet above the sea. The floor is now, as indicated in the picture, covered by a few inches of water at high tide, showing that a movement of depression is again in progress; and it is estimated that the land is now subsiding about one inch in four years. The upright position of the columns proves that these up and down movements of the temple have always been, like the present downward movement, gradual and quiet; and the neighboring volcanoes suggest that volcanic heat is the principal cause of the oscillations.

(2) *Sea-worn cliffs and caves above the reach of the waves at the present time.* The picture of Percé Rock, in the Gulf of St. Lawrence (**47**), shows, along the base of the cliff, the characteristic hollows and tunnels due to the action of the waves on the weaker portions of the rock; while precisely similar features are also clearly shown, especially near the right hand end of the picture, many feet above the present limit of wave action.

(3) *Elevated sea-margins or raised beaches.* When the elevation of the land is somewhat intermittent, beaches of sand and shingle, often with marine shells, and other traces of the shore-line, such as wave action, and corals

and shells sticking to the rocks, are left at different
levels, and the beaches often appear as distinct terraces
or level shelves fringing the coasts, as shown in the
view on the coast of Labrador (**2**).

(4) *The occurrence above sea-level of any rock which
shows evidence of having been formed in the sea.* The
Roxbury puddingstone (**41**), forming some of the higher
hills about Boston, is essentially a consolidated gravel-
beach, and proves a considerable elevation of the land
in this vicinity. The same is true of the slate and sandstone (**42**) so abundant in and around Boston. They are
hardened beaches and mud flats, and often show ripple-
marks or other special indications of their marine origin.
The evidence is still more striking when the rocks are
filled with marine shells (**43**) or other organisms. In
this way we learn that nearly all parts of the continents,
including those having the greatest altitude and most
remote from the shore, have been submerged beneath the
waters of the ocean, not once, but many times.

(5) *Submerged forests, peat-beds, etc.* At various
points along our coast, beds of peat, and the stumps or
roots of trees are found below sea-level, and yet in the
positions in which they grew, proving that the land is now
subsiding. The piece of wood (**45**) is from such a submerged forest on Cape Cod.

(6) *The deltas of rivers* often afford unmistakable evidences of gradual depression. Thus, the delta of the
Mississippi, for a depth of several hundred feet below the
present level of the sea, consists of layers of soil with
the stumps and roots of cypress trees in their original
positions and containing fresh-water or land shells alter-

nating with layers of clay and sand with marine shells.
These facts prove that the bottom of the delta-deposit
was once above sea-level, and that the subsidence has
been gradual and somewhat intermittent. The valley
of the Charles River, between Watertown and Charles-
town, is an equally good illustration. The Charles,
in this part of its course, flows over a considerable
thickness of sediments which include, below the present
level of Boston Harbor, layers of peat (**44**) with land
shells and stumps of coniferous trees. And these phenomena are repeated in the valleys of the Mystic, Nepon-
set and other streams of this region.

(7) *Fiords and drowned land-valleys generally.* The
fiords of Scandinavia, Alaska, and other regions, are
deep, narrow land-valleys which have been invaded by
the sea. This is proved by their sides preserving nearly
the same angle of slope below the surface of the water,
and thus giving a great depth near the shore, the depth
measuring the amount of subsidence. Soundings in the
vicinity of New York (see map **46**) show that the channel of the Hudson can be traced more than one hundred
miles beyond the present mouth of the river, and to a
depth of more than twenty-eight hundred feet.

(8) *The development of coral-islands and reefs.* The
reef-building corals are limited to a depth not exceeding
twenty fathoms; and yet in the Pacific and other oceans
the coral-limestone forming the islands and reefs often
has a thickness very much greater than this; indicating,
as will be more fully explained in a later section, that as
the reefs gained in hight the sea-bottom on which they
rest gradually subsided.

While the aqueous or sedimentary rocks of which the
continents are chiefly composed, teach us that the oscillations of the earth's crust have been very extensive and
wide-spread during the past geological ages, the map
(26) shows that these movements of the land and sea-
floor are still well-nigh universal. It is known that the
land areas colored purple on the map are rising at the
present time, and that those colored green are now subsiding.   It will be observed that these areas embrace the
greater part of the coast-lines of the globe, and the map
may be regarded as fairly accurate for the coastal regions,
but as almost entirely erroneous for the continental inte-
riors and ocean-beds, since it is certain that the movements are not limited to the shore-lines, and probable
that they are most important at the points farthest removed from the shore, which may be regarded, in many
cases, as the axis or fulcrum of the oscillations.

FORMATION OF MOUNTAINS.

If the model of Favre's experiment (24) showing the
plication of layers of clay by horizontal compression be
regarded as representing a large tract or section of the
earth's crust, it will not be difficult to understand how
mountains may be formed by the yielding or mashing up
of the crust under the tangential thrust, each irregular
ridge on the model representing a mountain range.   In
other words, mountains are formed by the collapse and
crushing up of the earth's crust along narrow but relatively weak zones, when the lateral pressure can no
longer be resisted.   The crust is shortened horizontally

and greatly thickened vertically, which explains the altitude of the mountains. It is in mountainous regions, almost exclusively, that the *reaction of the earth's interior upon its exterior* finally and permanently takes place. They are the culminating points of the play of the subterranean forces.

In considering the origin of mountains, we must distinguish carefully between mountain *forms* and mountain *structure*. Mountain-chains are the great theaters of erosion as they are of igneous action. "In all cases the erosion has been immense. In fact, as a rule, all that we see when we stand on a mountain-ridge — every peak and valley, every ridge and gorge, all that constitutes scenery, except the mere altitude — is wholly due to erosion."

A mountain-chain or system is produced solely, and the principal ranges of which it is composed, are produced chiefly, by the bulging of the crust by lateral pressure. But the minor longitudinal valleys and all the transverse valleys and cañons are always the work of erosion.

It follows from the foregoing that mountains when first elevated are much simpler externally than after they have been shaped by denudation. A recently elevated range would show its newness in its smoothly rounded outlines, unbroken crests, and general obtuseness. This contrast is well illustrated by the large relief maps in the west window space of Room B. We have reason to believe, however, that as a rule the elevation of mountains goes on so slowly that denudation destroys the original outlines as fast as they are formed.

The internal structure of mountains, it is almost unnecessary to add, is not influenced in any degree by the erosion; but is due to the yielding and crushing of the crust, with the exception of such structural features as may have existed in the rocks before the crushing and uplifting took place.

From our present position, then, it is clear that mountains are the result of two distinct processes :—

1. The crust yields, in obedience to the *interior* forces originating in the retreat of the central heat— *igneous* agencies, and the altitude, *general* outline, and most of the internal structure of the chain are produced.

2. The grand and simple bulge thus formed is modified or *sculptured* by *exterior* or *aqueous* agencies, developing most of the external structural or relief features of the chain.

These two processes or stages have been very appropriately named mountain *formation* and mountain *sculpture*. It is the mountain *formation*, only, that interests us now, our present object being, especially, to trace the operation of the subterranean or igneous agencies. Hence it is unnecessary to consider farther, at this point, the forms or external structure of mountains.

If the earth's crust were homogeneous, *i. e.*, equally strong in all parts, the crushing and wrinkling would, of course, be uniformly distributed. The surface would be generally roughened, as in a withered apple, but no distinct or dominant lines of elevation would be formed. Therefore, in a complete explanation of the origin of mountains we must account not only for a force adequate in direction and intensity, but also for the existence of the relatively weak zones in the crust.

To do this satisfactorily it is necessary to call attention first to two of the most important characteristics of mountain-ranges. (1) They are always composed of thick deposits of sedimentary or stratified rocks, and, usually, the size and other features of a mountain-system or range are proportional to the thickness of the sediments composing it; the thickness in most cases ranging from 20,000 to 60,000 feet. The section of the Alleghany Mountains (48) illustrates this important feature, and shows that, in consequence of the folding of the strata, a great part of the sediments is still below sea-level, and that their thickness exceeds the altitude of the mountains. The sedimentary rocks are, of course, formed below the level of the sea; and when they are squeezed horizontally and thickened vertically only a part of their mass is thus forced above sea-level.

(2) Mountain-ranges are usually, as shown in the introduction, near to and parallel with the shores of the ocean; if not of the existing ocean (Rocky Mountains, Andes, etc.), then of the ocean at the time the mountains were made (Alps, Ural Mountains, etc.).

These two features show that in their origin mountains are connected with the sea. For reasons which will be fully explained under the aqueous agencies, the thick accumulations of sediments of which mountains are formed can only be deposited in the marginal portions of the sea, following the coast-lines. Of course these belts of new sediments, at first, add little to the strength of the earth's crust; and before they become consolidated and strong they cause the interior heat of the earth to rise in and soften the underlying hard rocks. The zone of thick sediments parallel with the coast is thus gradually converted into a weak zone of the earth's crust, until finally it can no longer resist the growing horizontal pressure, but collapses, and the resulting mountain-range is added to the border of the continent.

The minor corrugations of the strata composing the earth's crust, the rock-folds or arches, so extensively developed in mountains, are, still more clearly than continents and mountain-ranges, the product of tangential compression. But the detailed explanations of their origin may be most conveniently considered in connection with the structural characteristics of the different kinds of folds, in the Guide to the Petrographic Collections. Faults, slaty cleavage, and other important structures incidental to the formation of mountains and resulting from the same great cause may be similarly treated. A vast amount and variety of geological structures are produced during, the growth of a mountain-range; and it is to the mountains that the student in almost every department of geology most naturally turns.

EARTHQUAKES.

The great earth-movements of elevation and subsidence and mountain-formation are not always perfectly smooth and steady; but they are accompanied by breaking, slipping and crushing of the rocks now and then; and, as a result of the shocks thus produced a swift vibratory movement or jar, which we know as an *earthquake*, spreads through the earth's crust.

Earthquakes are not only associated with movements of the earth's crust, but also with volcanic action; and the fact that volcanoes emit vast volumes of steam and other vapors with explosive violence makes it probable that earthquakes in volcanic regions are often due to the sudden expansion or condensation of steam; the earthquake, in this case, resembling the jar produced by the explosion of a keg of gunpowder buried in the earth.

An earthquake, whether due to an explosion or to the rupture or slipping of the rocks, is always, at its point of origin or focus, essentially a shock or blow. This primary shock or impulse causes a series of elastic vibrations or waves which spread out in all directions, as shown in the diagram on the chart (61), swinging the rocks to and fro through a few inches or a few feet and moving with a velocity of from 10 to 150 miles per minute. The point directly over the focus, where these vibrations first reach the surface, is called the epicentrum; and from this the shock spreads along the surface in an ever-widening circle. The direction of movement or shock is vertically upward at the epicentrum; but, as the radiating lines and arrows in the diagram show, the direction becomes more and more horizontal as the distance from the epicentrum increases. The tendency, therefore, is to throw vertically upward, or to crush, structures near the epicentrum, and to move horizontally or overthrow those remote from that point. It is by attention to these features that the position of the epicentrum and direction of movement are determined. The extremely sudden and violent to-and-fro movements or vibrations tend to form cracks or fissures in the earth, and in any structure resting upon it, at right angles to the direction of wave-movement, *i. e.*, at right angles to the radiating lines and arrows in the figure. Such earthquake-fractures are seen in the photograph. By observing where the perpendiculars to these fractures in different localities would meet if extended downwards in the earth, the position, depth and form of the earthquake focus are determined.

The distribution of the regions affected by earthquakes in recent times is indicated by the red shading on the small map of the world (62). The shading is darker in proportion to the force and frequency of the shocks. The black dots represent active volcanoes, and serve to show to what extent earthquakes are associated with volcanic phenomena.

VOLCANIC ACTION.

Under this head we properly consider all those agencies concerned in transferring *hot* materials from the earth's interior *to* or *towards* its surface. Every kind of volcanic activity requires a tube or opening leading up from deep-seated portions of the crust to or toward the opening.

This opening or conduit must usually, if not always, originate as a fissure or crack in the earth's crust. Eruptions may occur at all points or many points along an extended fissure, but often only at the widest part or at the intersection of two fissures, and thus a tube is formed.

The heated ejectamenta may be *solid* (stones, cinders, sand and dust), *liquid* (molten lava and water), or *gaseous* (steam and various acid vapors). Of these various products, the molten lava or liquid rock is by far the most characteristic and important.

A volcano is, then, fundamentally, a hole in the ground from which are ejected, in a highly heated condition, liquid and fragmental lavas and various gases or vapors. The accumulation of the lavas around the vent builds up the volcanic *cone*, with a cup-shaped crater at the top; but where the lava is sufficiently liquid it spreads out horizontally, forming a volcanic bed or sheet. While the lava which fails to reach the surface cools and solidifies in the fissures or conduits, formin *dikes*. Thus we have as the products of volcanic action, two great classes of igneous rocks :—(1) The volcanic rocks (**21-31**), which have cooled rapidly, under little pressure, and are usually light, porous, imperfectly crystalline and largely fragmental ; and (2) the plutonic or dike rocks (**41-44**), which have cooled slowly and under great pressure, and are, as a rule, dense and crystalline.

The aspects of a typical volcano in a nearly quiescent state, and during a great eruption, are shown in the views of Vesuvius (**1**) ; while the accompanying diagram illustrates the general relations of the volcanic cone and conduit or fissure to the earth's crust, which is repre- sented as composed of stratified or sedimentary rocks in the upper part and massive or igneous rocks in the lower part.

Volcanic eruptions may be divided into two great types, viz., the *quiet* and the *explosive*. In the first, the lava is very fluid and flows out quietly from the central crater or through lateral fissures, with few explosions and little fragmental material; while in the second the material is largely blown out in the form of dust and fragments by explosions of steam, accom- panied by violent earthquakes. The Hawaiian volcanoes are perhaps the best examples of the first class, and the Javanese volcanoes of the second class.

The form of a volcanic cone evidently depends upon the character of the eruptions and the fluidity of the lava, and especially upon the relative proportions of lava and fragmental materials. Thus, Etna (**83**, section 3) is built up chiefly of rather fluid lava, which spreads far from the vent, so that it is a broad low mound with very gentle slopes ; while Vesuvius (**87**), consisting largely of fragments and dust formed by the violent explosions accompanying its eruptions, and which fall mainly in the immediate vicinity of the crater, presents the more conical form and steeper slopes shown in the photographs already referred to.

Every important volcanic cone is the product of intermittent eruptions during a long period of years, and consequently must consist of many successive layers of lava and of fragments or volcanic tuff, all sloping from the central orifice or crater. This normal structure of a cone is shown in the diagram accompanying the photograph of Monte Somma and the summit of Vesuvius (**2**). The summit of Vesuvius, shows the recently formed cinder-cone, the steam and acid vapors escaping from the lava, and, on the left, the contorted, rope-like structure characteristic of most lava-streams. This is also shown, together with the steam holes or vesicles formed by the expansion of steam in the liquid lava, in the large specimen (**30**), which is from the flow (1872) represented in the general view of Vesuvius in a state of active eruption (**1**).

Eruptions usually occur not only from the summit-crater of the volcano, but also often from lateral fissures. By the enormous hydrostatic pressure exerted by the liquid lava in the main crater and its conduit or throat the cone is burst open, and fissures are formed radiating from the crater. These cracks are filled with lava, forming dikes which intersect the successive layers of lava, so as to bind them together and strengthen the

volcano as a whole. Through these fissures, when first
formed, the principal streams of lava often pass; and
upon them subordinate craters and cones are finally
formed. These smaller cones, appearing like pimples
on the slopes of the main cone, are called monticules.
There are about 600 monticules on the slopes of Etna,
only a few of the larger ones being shown in the model
of that volcano (**83**, section 3). From one of these
on the south side of Etna descended the great lava-
flow of 1669, which overwhelmed fourteen, towns and
villages and flowed into the sea.

The great depression in the eastern side of Etna, called the
Val del Bove, and shown in the model, is probably due to a subsidence into a subterranean cavity formed by the enormous out-
flows of lava which have built up this gigantic cone; although
it may possibly be due to some great explosion which has
blown out the whole side of the volcano.

The immense amount of fragmental material ejected
during some explosive eruptions with such force, or of
such fineness as to be carried beyond the limits of the
volcano itself, is clearly illustrated by the buried city of
Pompeii. The exhumed ruins of this Roman town are
shown in the photograph (**48**), with Vesuvius, the cause
of the disaster, in the back ground. The city is several
miles from the base of the volcano, and its site has not
been covered by the molten lavas within historic times;
but during the first historic eruption of Vesuvius, in the
year 79, the whole floor and part of the wall of the ancient crater were blown away and Pompeii and the surrounding country were deeply buried under the volcanic
dust and fragments of pumice.

The relief-map or model of Vesuvius (87) shows clearly how large a part of the rim or wall of the ancient crater was blown away at this time; and, also, in the darker color of the western and southern slopes, how extensive and wide-spread the subsequent out-flows of lava have been. The main, central cone of Vesuvius, some 1800 feet high, had no existence before the great eruption of 79, but is wholly made up of the dust and fragments and liquid lava thrown out during that and later eruptions.

Some of the principal events in the history of an active volcano, and the influence of its lava-streams upon the drainage and general topography of the country, are illustrated by the small models (81–82, section 3). The details of these are, however, sufficiently explained on the descriptive labels.

The small map of the world (32) shows the general distribution of volcanic phenomena at the present time. Each red dot represents an important cone or group of cones. The important feature of the distribution is that the active volcanoes of the globe, with a very few exceptions, are all on islands or on the margins of the continents. It is especially noticeable that the Pacific Ocean is bordered by a nearly continuous chain of volcanoes, from Patagonia to Behring's Strait, and from Behring's Strait to New Zealand.

Geologists have been able to show with much probability that this association of volcanoes with the coast-lines and the sea has existed during the whole of geological time. Thus, wherever volcanic rocks are found in the interior of the continents we have reason to believe that at the time of their eruption they were in or near the sea; the shore-line having subsequently receded in consequence of the constant vertical oscillations of the earth's crust.

Compared with the whole of geological history, volcanoes are short-lived. Their eruptions culminate, and then become fewer and weaker, and finally cease altogether. Erosion then rapidly wears away the more exposed portions of the volcanic mass; and by the time the shore-line has receded so as to leave it in the interior of the continent, the volcano is permanently extinct and has lost its characteristic form.

Geologists are still in doubt as to the cause of the association of active craters with the sea, and it remains to be determined how far, if at all, volcanic action depends upon access of the waters of the ocean to the highly heated subterranean regions. In fact, the whole question as to the exact cause of volcanic action is still unsettled.

Volcanic phenomena may be classed as primary, where attended by eruptions of liquid or solid lava, and secondary, where no lava is ejected, but only hot water and heated vapors and gases. The secondary accompany the primary phenomena and usually continue long after the outflows of lava have ceased and the volcano is otherwise extinct. Or they may exist quite independently of outflows of lava, where heat has been generated through rock-crushing by horizontal pressure, as already explained. But thick flows of lava remain hot in the interior for an incalculable time; and the secondary phenomena, including geysers, hot springs and solfataras, are due chiefly to the percolation of water through these heated masses. Solfataras exist where heated vapors, and especially sulphurous vapors, escape from the rocks. Deposits of sulphur, sal ammoniac, and other minerals are often formed about the vents. The specimen of sulphur (45) is from the Solfatara near Naples, Italy.

The escape of hot water gives rise to thermal springs; and when the water is very hot, so that violent eruptions of water take place periodically, the hot spring becomes a geyser. The thermal water, under the great pressure existing at considerable depths, decomposes the heated rocks through which it flows and large amounts of various minerals go into solution; and then, as the water escapes toward the surface, and its temperature and pressure are diminished, the dissolved minerals are largely deposited on the walls of the fissures through which it flows, until, finally, the fissures are completely filled and the water forced to seek a new outlet. In this way the various kinds of mineral veins are formed; and it has been conclusively proved that metalliferous and other veins are being made in this way at the present time, in California, Nevada, and other localities.

Where the thermal water issues on the surface, a farther portion of the dissolved minerals is often deposited, building up very characteristic masses which are called tufas. The substances most commonly deposited in this way are silica and carbonate of lime, forming siliceous tufa (46) and calcareous tufa (47). See also the photograph (49).

METAMORPHISM.

We have seen that the interior heat of the earth is slowly but steadily diminishing, and that this secular

cooling of the earth's interior probably contributes in an
important degree to the development of the enormous
tangential pressure known to exist in the earth's crust;
also that this pressure is the chief agent in (1) determining the oscillations of the earth's crust and the distribution of land and sea, and (2) the formation of mountains, with all the attendant phenomena of rock-folds,
faults, earthquakes, etc. Again, we have observed that
the loss of the subterranean heat by conduction is supple-
mented in an important measure by volcanic action, or
the transfer of heated matter, especially molten lava and
water, from the earth's interior to the surface, giving
rise, in its primary stages, to the great class of eruptive
or igneous rocks, and, in its secondary stages, where heat
coöperates with water, to many kinds of vein-rocks and
tufas.

But the geological work of the subterranean heat does
not end here; for it invades the stratified or sedimentary
rocks, formed by the action of water alone on the earth's
surface, and causes various important changes in composition and structure. These aqueous rocks are, for the
most part, when first formed, deposits of gravel, sand,
mud, shells, etc.; imperfectly consolidated and wholly uncrystalline. After their formation, as will be more fully
explained elsewhere, they are consolidated and otherwise
changed by heat, pressure and chemical action. All
these changes are properly metamorphic; although this
term is very generally restricted to the processes resulting
in the crystallization of the sedimentary rocks. The exact
causes of metamorphism are not definitely known in many
cases, but it is certain that heat is usually a chief agent,

although coöperating, as a rule, with water, pressure, and chemical action.

Local metamorphism or alteration of sedimentary rocks is observed chiefly in the neighborhood of masses of igneous rocks; and in nearly all such instances the volcanic heat is plainly the principal agent. Thus, the native coke (**81**) has been formed, as coke usually is, by the action of heat upon bituminous coal; the heat coming in this instance from a dike of trap which has broken through a coal-bed. The alteration is most marked next to the dike, and dies out at a distance of a few feet or yards. Under very similar conditions chalk has been changed locally to the hard, dense limestone or marble shown in the next specimen (**82**). The specimen of black slate from the vicinity of large dikes of diabase at East Point, Nahant (**83**), has been very thoroughly baked and indurated by the heat of the dikes; and the expansion of vapors due to the heat has developed lenticular cavities along certain planes in the rock, in which epidote and other minerals resulting from the metamorphism have crystallized. A still more perfect example of crystallization is seen in the slate from the quarries in Somerville (**84**). The pyrite crystals are there observed in the slate only within a few inches or feet of the trap dikes. The next specimen (**85**) is slate from near the contact with the granite in Quincy. A crystalline micaceous mineral has been largely developed in it, apparently at the time when it was pervaded by the heat of the molten granite. Metamorphism is not limited to the sedimentary rocks; but the eruptive rocks are also subject to extensive morphologic and chemical changes; and through the combined action of mechanical and chemical forces they frequently come to resemble very closely some of the crystalline schists and gneisses **(86)**, and other sedimentary rocks.

SUPERFICIAL OR AQUEOUS AGENCIES.

When we think of the ocean with its waves, tides, and currents, of the winds, and of the rain and snow, and the vast net-work of rivers to which they give rise, we realize that the energy or force manifested upon the earth's surface resides chiefly in the air and water — in the earth's fluid envelop and not in its solid crust. And it is an easy matter to show that, with the exception of the tidal waves and currents, which of course are due chiefly to the attraction of the moon, nearly all this energy is merely the transformed heat of the sun. Now the air and water are two great geological agencies, and therefore the geological effects which they produce are traceable back to the sun. Organic matter is another important geological agent ; but all are familiar with the generalization that connects the energy exhibited by every form of life with the sun.

Of this trio of geological agencies operating upon the earth's surface and vitalized by the sun — *water, air,* and *organic matter* — the water is by far the most important, and so it is common to call these collectively the aqueous agencies. The aqueous agencies include, on one side, *air* and *water*, or *inorganic* agencies ; and, on the other, *animals* and *plants*, or *organic* agencies.

The collection is arranged to illustrate, as well as may be, the operation of these two classes of agencies, beginning with the air and water. But it is necessary, first, to note that nearly all the geological work of the superficial or aqueous agencies may be considered under the two general heads of erosion, or the wearing away of the rocks of the land, and deposition, or the formation in the sea from the material worn off from the land of the various kinds of sedimentary rocks ; and that each of these processes may be chiefly chemical or chiefly mechanical in its nature.

AIR AND WATER, OR INORGANIC AGENCIES.

CHEMICAL EROSION.

The illustration begins with a local example (**41–44**). The first specimen (**41**) is a sound, fresh piece of the rather common rock, diabase ; and those who are acquainted with minerals will recognize that the light-colored grains in the rock are feldspar, and the dark, augite. This specimen came from a depth in the quarry, and has not been exposed to the action of the weather.

The second specimen (**42**) differs from the first, apparently, as much as possible ; and yet, except in being somewhat finer grained, it was originally of precisely similar composition and appearance. In fact, it is a portion of the same rock, but a *weathered* portion. In this we can no longer recognize the feldspar and augite as such, but both these minerals are very much changed, while in the place of a strong, hard rock we have an incoherent, friable mass, which is, externally at least, easily crushed to powder; and with the next step in the weathering, as we may readily observe in the natural ledges, the rock is completely disintegrated, forming a loose earth or soil. The third and fourth specimens (43–44) are two examples of such natural powders; and by washing these (especially the finer one) with water, we can prove that they consist of an impalpable substance which we may call clay, and angular grains which we may call sand. The sand-grains are really portions of the feldspar not yet entirely changed to clay.

Thus we learn that the result of the exposures of this hard rock to the weather is that it is reduced to the condition of sand and clay. What we mean especially by the weather are *moisture* and certain constituents of the air, particularly *carbon dioxide*. The action of the weather on the rocks is mainly chemical. With a very few exceptions, the principal minerals of which rocks are composed, such as feldspar, hornblende, augite, and mica, are silicates, *i. e.*, consist of silicic acid or silica combined with various bases, especially aluminum, magnesium, iron, calcium, potassium, and sodium.

Now the silica does not hold all these bases with equal strength; but carbon dioxide, in the presence of moisture, is able to take the sodium, potassium, calcium, and magnesium away from the silica in the form of carbonates, which, being soluble, are carried away by the rain-water. The silicate of aluminum, with more or less iron, takes on water at the same time, and remains behind as a soft, impalpable powder, which is kaolin or common clay.

In the case of the diabase, continued exposure to the weather would reduce the whole mass to clay. But other rocks contain grains of quartz, a hard mineral which cannot be decomposed, and it always forms sand. Certain classes of rocks, too, such as the limestones and some iron-ores, are completely dissolved by water holding carbon dioxide in solution, and nothing is left to form soil, except usually a small proportion of insoluble impurities like sand or clay. Thus, the soil on the upper side of the specimen of limestone (**45**) is the insoluble impurity or clay contained in a considerable thickness of the limestone which has been removed by solution.

It is interesting to notice how these agents of decay get at the rocks. Neither water nor air can penetrate the solid rock or mineral to any considerable extent, so that practically the action is limited to surfaces, and whatever multiplies surfaces must favor decomposition.

First, we have the upper surface of the rock where it is bare, but more especially where it is covered with soil, for there it is always wet. All rocks are naturally divided by joints into blocks which are frequently more or less regular, and often of quite small size. Water and air penetrate into these cracks and decompose the surfaces of the blocks, and thus the field of their operations is enormously extended. These rock-blocks sometimes show very beautifully the progress of the decomposing agents from the outside inward by concentric layers or shells of rotten material, which, in the larger blocks, often envelop a nucleus of the unaltered rock (**46**).

It is interesting to observe, too, that these concentric lines of decay cut off the angles of the original blocks, so that the undecomposed nucleus, when it is found, is approximately spherical instead of angular.

In the rocks also we find many imperfect joints and minute cracks. In cold countries these are extended and widened by the expansive power of freezing water, and thus the surfaces of decomposition become constantly greater.

Nearly all rocks suffer this chemical decomposition when exposed to the weather, but in some the decay goes on much faster than in others. Diabase is one of the rocks which decay most readily; while granite (**48**) is, among common rocks, one of those that resist decay most effectually.

The specimen from Louisburg Square (**47**) shows, however, how rapidly even the granites may yield under favorable conditions, when frost coöperates with the chemical agents.

The caverns which are so large and numerous in most limestone countries are splendid examples of the solvent action of meteoric waters, being formed entirely by the dissolving out of the limestone by the water circulating through the joint cracks. The process must go on with extreme slowness at first, when the joints are narrow, and more rapidly as they are widened and more water is admitted. We get some idea, too, of the magnitude of the results accomplished by these silent and unobtrusive agencies when we reflect that almost all the loose earth and soil covering the solid rocks are simply the insoluble residue which carbon dioxide and water cannot remove.

In low latitudes, where a warm climate accelerates the decay of the rocks, the soil is usually from 50 to 300 feet deep.

The remaining specimens on the third shelf (**49–52**) are a very perfect illustration, from the vicinity of Washington, of the characteristic sedentary or residuary soil of the South, showing four stages in the transformation of a hard, gray, micaceous gneiss into a bright red clay soil, and matching the diabase series (**41–44**) from this vicinity.

MECHANICAL EROSION — ON THE EDGE OF THE LAND.

Whoever has been on the shore must have noticed that the sand along the water's edge is kept in constant motion by the ebb and flow of the surf. Where the beach is composed of gravel or shingle the motion is evident to the ear as well as the eye ; and when the surf is strong, the rattling and grinding of the pebbles as they are rolled up and down the beach develops into a roar.

The constant shifting of the grains of sand, pebbles, and stones is, of course, attended by innumerable collisions, which are the cause of the noise. Now it is practically impossible, as one may easily prove by experiment, to knock or rub two pieces of stone together, at least so as to produce much noise, without abrading their surfaces ; small particles are detached, and sand and dust are formed.

That this abrasion actually occurs in the case of the moving sand is most beautifully shown by the sand-blast. We are to conclude, then, that every time a pebble, large or small, is rolled up and down the beach it becomes smaller, and some sand and dust or clay are formed which are carried off by the water.

But what are the pebbles originally ? This question is not difficult. A little observation on the beach shows that the pebbles are not all equally round and smooth, but many are more or less angular ; and we soon see that it is possible to select a series showing all gradations between the most perfectly rounded forms and angular fragments of rock that are only slightly abraded on the corners. The three principal members of such a series are shown in the specimens from the beach on Marble-head Neck (25) ; but equally instructive specimens can be obtained at many other points on our coast. It is also observable that the well-rounded pebbles are much smaller on the average than the angular blocks.

From these facts we draw the legitimate inference that the pebbles were all originally angular, and that the same abrasion which diminishes their size makes them round and smooth. A little reflection, too, shows that the rounding of the angular fragments is a natural and nec-essary result of their mutual collisions ; for the angles are at the same time their weakest and most exposed points, and must wear off faster than the flat or concave surfaces.

The next specimen (29) shows how the softer parts of the rock-fragments are worn away more rapidly than the hard parts, the latter forming the salient portions of the rounded pebbles. Typical examples of well-rounded pebbles are also seen in the specimens from the beaches at Newport (27-28). Whether the forms are circular, spherical or oblong depends mainly upon the general shape of the original fragments, i. e., upon the way in which the parent ledge or cliff breaks up.

The formation of rounded pebbles by abrasion is very beautifully illustrated by fragments of brick, coal, or glass (26) which have been worn smooth and round by being rolled about on the beach ; for it is certain in the case of such specimens that they were originally rough and angular.

Having traced each pebble back to a larger angular rock-fragment, the question arises, Whence come these angular blocks? Behind the gravel-beach, or at its end, there is usually a cliff of rocks. As we approach this it is distinctly observable that the angular pebbles are more numerous, larger, and more angular; and a little observation shows that these are simply the blocks produced by jointing, and that the cliff is entirely composed of them. In other words, the cliff is a mass of natural masonry, which chemical agencies, the frost, and the sea are gradually disintegrating and removing. As soon as the blocks are brought within reach of the surf their mutual collisions make them rounder and smaller; and small, round pebbles, sand, and clay are the final result.

Where the waves can drive the shingle directly against the base of the cliff, this is gradually ground away in the same manner as the loose stones themselves, sometimes forming a cavern of considerable depth, but always leaving a smooth, hard surface, which is very characteristic, and contrasts strongly with the upper portion of the cliff, which is acted on only by the rain and frost. A good example of such a pebble-carved cliff may be seen behind the beach on the seaward side of Marblehead Neck.

The sea acts within very narrow limits vertically, a few feet or a few yards at most; but the coast-lines of the globe (including inland lakes and seas) have an aggregate length of more than 150,000 miles. Hence it is easy to see that the amount of solid rock ground to powder in the mill of the ocean-beach annually must be very considerable. Waves, cutting ever at the shore-line only, act like an horizontal saw. The receding shore-cliff, therefore, leaves behind it an ever-increasing submarine platform which marks the amount of recession. See the photograph (**1**). The rate of this marine erosion and the form of the coast depend upon both the force of the waves and currents and the nature and structure of the rocks. The irregular coast of this region is due mainly to the fact that, on account of differences in composition and structure, the rocks vary greatly in the resistance which they offer to the action of the waves. The waves and currents not only have great power to break up and wear away the land, but also to transport the debris resulting from their action.

It is swept along the coast into some sheltered bay or carried out by the ebbing tide into deep water. These principles are well illustrated by the promontories and islands of Boston Harbor; and by the models (**21–24**), which should be studied in order, beginning on the left.

MECHANICAL EROSION — ON THE SURFACE OF THE LAND.

It is a familiar fact that after heavy rains the roadside rills carry along much sand and clay (which we know have been produced by the previous action of chemical forces), and also frequently small pebbles or gravel. It is easy to show that in all important respects the rills differ in size only from brooks and rivers; and the former afford us fine models of the systems of valleys worn out during the lapse of ages by rivers. The turbidity of rivers is often very evident, and in shallow streams we can sometimes see the pebbles rolled along by the current.

Now here, just as on the beach, the collisions of rock-fragments are attended by mutual abrasion, sand and clay are formed, and the fragments become smaller and rounder. Our series of pebbles from the beach might be matched perfectly in the river-gravel. In mountain streams especially we may often observe that pebbles of a particular kind of rock become more numerous, larger, and more angular as we proceed up stream, until we reach the solid ledge from which they were derived, showing the same gradation as the beach pebbles when followed back to the parent cliff.

The pebbles, however, not only grind each other, but also the solid rocks which form the bed of the streams in many places, and these are gradually worn away. When the rocky bed is uneven and the current is swift, pebbles collect in hollows where eddies are formed, by which they are kept whirling and turning, and the hollow is deepened to a pot-hole, while the pebbles, the river's tools, are worn out at the same time.

By these observations we learn not only that running water carries away sand and clay already formed, but that it also has great power of grinding down hard rocks to sand and clay. The ocean is the common goal of nearly all rivers; and therefore the constant tendency of the rain falling upon the land is to break up the rocks by chemical and mechanical action and transport the debris into the sea.

The erosive power of water is most easily studied in ravines, gorges, and cañons, and especially in waterfalls. Every waterfall is slowly moving up stream, cutting its way back through the rocks over which the stream pours. Hence, while the gorge is deepened by the action of the stream on its bottom, it is also lengthened at the upper end by the recession of the escarpment. In many cases, as at Niagara, and in the examples illustrated by the photograph (2) and the model (61), the gorge is evidently the product chiefly of the cascade. By carefully observing the present rate of recession of a waterfall, the time required for the excavation of its gorge can be approximately determined; and in this way some of our most reliable estimates of the length of geological time have been reached. The different time-estimates for Niagara range from 5,000 to 40,000 years. But all our estimates are in harmony on the main points; viz., that erosion is a slow process and that geological time is immensely long.

The principal factors in stream-erosion are : (1) Hight above sea-level ; for, of course, streams cannot cut below the level of the sea, and altitude is required to give them force or erosive power. (2) Amount of water. (3) Character and structure of the rocks. The valley shown in the next model (62) is characteristic of a moderately elevated country, the surface of which is exposed to the action of frost and a generous rainfall. The stream cuts down its bed, while the rain and frost wear down the whole face of the country, rounding off or bevelling the edges of the valleys. The strata are represented as horizontal, and such inequalities as appear in the slopes are due to the unequal hardness of the layers of rock. The companion model (63) and photograph of the Grand Cañon of the Colorado (53), on the other hand, show the deep, narrow valleys or cañons characteristic of a more elevated and much drier country. Cañons, as explained on the labels, have their typical development where arid plateaus, like those of Utah, Colorado, New Mexico, and Arizona, are traversed by rivers deriving their water from distant mountains. The rivers, under these conditions, cut deep narrow trenches or cañons across the plateau ; for the walls of the cañons, being but little exposed to the action of rain and frost, retain their upright form. In this connection, the visitor should examine the large relief map of the Grand Cañon of the Colorado and the High Plateau in the west window-space of Room B.

These stupendous examples of erosion are very impressive. It is necessary to remember, however, that the erosive power of streams is in most cases far inferior to that of rain and frost, because the latter act upon such vastly larger areas. But the transporting power of running water is very great ; and carrying the products of chemical and mechanical erosion to the sea is, after all, the chief office of rivers, geologically considered.

The transporting power of a current varies as the sixth power of the velocity, which means that doubling the velocity increases the transporting power sixty-four times ; and since the velocity depends upon the rate of fall or descent of the stream bed, it is clear that transportation as well as erosion will be most effective in the upper or mountainous parts of the rivers. This is well illustrated by the large photograph (81) of a mountain valley in Norway. It is evident that during the period of high water the floor of the valley is swept by a torrent carrying immense amounts of debris, the source of which is sufficiently manifest in the talus of shattered rocks heaped against the base of the cliffs and produced by the quiet action of rain and frost.

When, from any cause, such as a diminution of water or of slope or a broadening of its channel, the velocity of a stream is diminished, a large part of the debris which it carried or rolled along is deposited. The flood-plains of its lower courses (photograph 54) and the delta at its mouth are formed of these deposits. The photograph (30) of the detrital cone at Silvaplana is a fine example of the fan-shaped deposit of debris, or delta, formed where a mountain torrent enters the quiet waters of a lake. The gradual formation of a delta at the mouth of a river flowing into the sea is illustrated by two models (64–65).

Rivers are continually uniting to form larger and larger streams; and thus the drainage of a wide area sometimes, as in the case of the Mississippi Valley, reaches the sea through a single mouth. By careful measurements made at the mouth of the Mississippi it has been shown that the 20,000,000,000,000 cubic feet of water discharged into the Gulf of Mexico annually carries with it no less than 7,500,000,000 cubic feet of sand, clay, and dissolved mineral matter; and this, spread over the whole basin of the Mississippi, would form a layer a little more than $\frac{1}{5,000}$ of a foot in thickness. So that we may conclude that the surface of the continent is being cut down on the average about *one foot* in *five thousand years*.

We turn next to the very important geological action of water in the solid state, as in glaciers and icebergs. The moisture precipitated from the atmosphere, and falling as rain, makes ordinary rivers; but falling in the form of snow in cold or elevated regions, where more snow falls than is melted, the excess accumulates and is gradually compacted to ice, which, like water, yields to the enormous pressure of its own mass and flows toward lower levels. When the ice-river or glacier reaches the sea it breaks off in huge blocks, which float away as icebergs. Moving ice, like moving water, is a powerful agent of erosion; and the glacial marks or scratches observable upon the ledges everywhere in the Northern States and Canada attest the magnitude of the ice-action at a comparatively recent period. The photograph (1) of an elevated Alpine valley shows the birthplace or source of a glacier in the broad and deep accumulation of granular, half compacted snow or *névé*. This moves slowly but steadily down the steep slopes and is soon changed by pressure and repeated thawing and freezing to true glacier ice.

The next photograph (26) shows how the nearly perfect ice is broken and crushed in its movement over a steeper and more irregular slope. While the photograph of the Viesch Glacier (2) is a general view of a long, well-defined, typical, Alpine glacier from its source in the field of *névé* nearly to its lower limits, where the wasting of the ice by the sun and air just balances its downward movement. It shows how perfectly the ice-stream adapts itself to the serpentine form of the valley; and also the wonderful way in which it is fissured or crevassed in passing over and around the convex surfaces. The *lateral* moraines are distinctly shown along either edge of the glacier. These are bands or ridges of debris which has fallen upon the ice from the crumbling cliffs, or has been torn by the glacier itself from the base of the cliffs. Where two glaciers unite, as in the upper end of this view, the two adjacent lateral moraines are combined to form one *medial* moraine, which is shown here as a dark, serpentine ridge of debris extending down the middle of the glacier from the point of union to its lower limit.

An exceptionally fine development of medial moraines is afforded by the great Gorner Glacier and its numerous tributary glaciers descending from the Monte Rosa and Matterhorn range, as seen in the panorama from the Gorner Grat (48).

Model (21) shows, in some respects, a more complete system of Alpine glaciers, from the lofty snow-fields and *névé*, down along the glacier proper with its medial and lateral moraines, to the terminal moraine and the stream which issues from the lower end of the glacier. The relief map of the Mt. Blanc Range in section 47, Room B, is also of particular interest in this connection.

All the material carried on, in or under the ice is, of course, dropped at the lower end of the glacier and contributes to the formation of the terminal moraine. This immense accumulation of debris, which is essentially the delta of the glacial river, is pushed by the advance of the glacier into a steep, crescentic ridge.

The most important geological work accomplished by glaciers is the erosion of the rocks over which they move and the transportation of the debris in the different kinds of moraines. In existing glaciers, the transportation is very obvious, and the wearing or erosion also, where a temporary retreat of the ice exposes the rocky bed over which it has recently moved. For the rocks in such positions are always smooth and rounded and marked by grooves and scratches running in the direction in which the ice has moved   The block of fine hard granite (82,

section 5) has been ground down and almost polished by
glacial action.   The stones or bowlders dragged along by the
ice over the ledges are also smoothed and striated, as witness
the glaciated bowlder of slate from East Boston on the right
hand side of the stairs.   It is probable, however, that the
erosive action of glaciers has been over-estimated, and that the
principal effects are due to the streams of water which run under the ice and roll along stones and sand in the usual manner.
That the great ice-sheet moved for hundreds of miles from
north to south across the country is proved by the occasional
occurrence in the glacial detritus or drift of bowlders which are
very far removed from their nearest outcrops to the northw{
This evidence is especially conclusive when the rock is {
unique character and its outcrop limited.   Thus the red jasper
conglomerate represented by the bowlder near the stairs is
found in place only in a small area north of Lake Huron, but
bowlders precisely similar to this are scattered in the drift as
far south as the Ohio River, nearly six hundred miles from their
point of origin.

    The bowlder of granite from the top of Mt. Washington is
wholly unlike any rock *in situ* on that mountain; but twelve
miles to the north-northwest, on Cherry Mountain, we find precisely the same kind of granite, at a level 3,000 feet lower,
showing that the ice-sheet passed from Cherry Mountain up
over the summit of Mt. Washington, the highest point in New
England.

    In the polar regions, glaciers are developed on a far
grander scale ; for they are not limited to elevated
mountain valleys but cover the whole face of the country
to a depth of hundreds or thousands of feet.   The Antarctic Continent appears to be almost completely buried
under a thick sheet of ice, and recent explorations have
shown that Greenland is a vast *mer de glace*, from which
enormous glaciers descend on every side through great

fiords to the sea. The photograph (49) shows one of the great Greenland glaciers entering the sea, where the ice is breaking off and forming icebergs.

During the comparatively recent geological period known as the glacial epoch or great ice age the conditions which are now limited to Alpine districts and high latitudes, appear to have extended much nearer the equator, nearly all of eastern North America down to latitude 40° having been covered by a continuous sheet of ice thousands of feet thick. Its existence and the direction of its movement are plainly indicated everywhere in this region by the polishing and grooving of the ledges and the great masses of drift or moraine material of different kinds, which we find now almost as they were left by the ice-sheet. The principal phenomena of the continental glacier or ice-sheet in its slow progress over a comparatively level country and gradual retreat as the climate became milder toward the end of the glacial epoch are illustrated by the series of four models (22-25). The necessary explanations are given on the labels.

We have already noticed incidentally the powerful disintegrating action of water where it freezes in the joints and pores of the rocks; and it is probable that it thus facilitates the destruction of the rocks in cold countries nearly as much as the higher temperature and greater rainfall do in warm countries.

Air, especially as an agent of transportation, acts to a limited extent independently of water. The dunes or sand-hills north of Cape Ann and on Cape Cod and other parts of our coast are excellent illustrations of the extent to which the dry sand on the upper edge of the beaches may be carried inland and piled

up by the wind. While the windows of houses in such districts, and the sand-blast used for etching glass, show that the
wind-borne sand has considerable erosive power. Finer material or dust, especially of volcanic origin, is carried for long
distances by the air currents, and spread over large areas of the
earth's surface. But the principal office of the wind is to im-
part its energy to the water, which is the great and efficient
agent of erosion; and a summary of this section can be given in
a few words :—

The heat of the sun lifts the water from the ocean in the form
of invisible vapor, and it also warms the air unequally, creating
currents or winds. The winds cause waves and currents which
are ever gnawing at the coast-lines of the globe and wearing
away the land; and they also bear the aqueous vapor over the
land, where it descends in the form of rain and snow. The
rain-water, as soon as it falls, begins to flow towards lower
levels and the sea, in streams which by their confluence become
ever larger and fewer, giving in each hydrographic basin an insensible gradation from a myriad tiny rivulets to one large
river. At the river's mouth the cycle of the circulation is completed. The rain-water flows through as well as over the rocks
of the land, and, aided to a certain extent by the air, it acts
both mechanically and chemically upon the rocks, disintegrating
and decomposing them, and forming loose material or soil. To
a limited extent this detritus is formed and transported by ice
in the forms of glaciers and icebergs, and ny the air alone; but
running water is throughout the chief agent. Of the detritus
swept along by streams, a small portion may be temporarily or
permanently deposited in quiet portions of their channels, but
the greater part ultimately reaches the sea. Thus the land is
wasted away and valleys are formed. which are a measure of
the erosion.

Our observations up to this point show us, then, that
erosion, by which we mean the breaking up by chemical

and mechanical action of the rocks of the land and the transportation of the debris into the sea, is one great result accomplished by the inorganic aqueous agencies.

The products of erosion carried into the sea from the land include two distinct portions: gravel, sand, and clay, which are held in suspension by the streams and waves; and carbonate of lime and other salts, which are held in solution.

MECHANICAL DEPOSITION.

We have next to consider what becomes of all this vast amount of clay, sand, and gravel, or matter in suspension, after it is washed into the ocean; although, unfortunately, this subject is one that does not admit of satisfactory illustrations within the narrow limits of the collection.

By taking up a glass of turbid water from any roadside rill, after a heavy rain, and observing that as soon as the water is undisturbed the sand and clay begin to settle, we learn that the solid matter is not held in suspension long after being washed into the sea, for the water otherwise would, in the course of time, become turbid for long distances from shore; and it is a well-known fact that the sea-water is usually clear and free from sensible turbidity close along the shore, and even near the mouths of large rivers, while at a distance of only 50 or 100 miles we find the transparency of the central ocean.

Putting these facts together, we see that the ocean, notwithstanding the ceaseless and often violent undulations of its surface, must be as a whole a vast body of still water; and to the reflecting mind the almost perfect tranquillity of the ocean is one of its most impressive features; for it is in striking contrast, in this respect, with the more mobile aerial ocean above it.

The rapid and complete precipitation of the finest mechanical detritus or clay in the sea is made still clearer by a simple experiment. Into each of two bottles put a small amount of fine clay; fill one bottle with fresh water and the other with salt water; bring the clay in each bottle into perfect suspension by violent agitation; and then allow the bottles to remain undisturbed for several hours or days. The salt water will become quite clear by the complete settling of the clay, while the fresh water remains distinctly turbid, showing that the salt favors the rapid deposition of the clay. The fact is, the clay is not held in suspension wholly by the *motion* of the water; but, just as in the case of dust in the atmosphere, a small portion of the medium is condensed around or adheres to each solid particle, *i. e.*, each clay particle in this experiment has an atmosphere of water which moves with it and buoys it up. Now the effect of the salt is to diminish their atmospheres, and consequently their buoyancy. The diminished adhesion of the salt water is well shown by the smaller drops which it forms on a glass rod.

The geological importance of this principle is very great; for it is undoubtedly largely to the saltness of the sea that we owe its transparency, and the fact that the fine, clayey sediment from the land, like the coarse, is deposited near the shore.

We have got hold, now, of two facts of great geological importance : (1) The debris washed off the land by waves and rivers into the still waters of the ocean very soon settles to the bottom ; and (2) it nearly all settles on that part of the ocean-floor near the land. And now we have in view the second great office of the inorganic aqueous agencies, — deposition, the counterpart or complement of erosion.   The land is the theater of erosion and the sea of deposition ; the rocks which are constantly wasting away on the former are as constantly renewed in the latter.   We will now examine the process of deposition a little more closely.   If a mixture of gravel, coarse sand, fine sand, and clay is thrown into quiet water, they will be deposited in the following order : the gravel falls to the bottom almost instantly, followed quickly by the coarse sand and very soon afterward by the fine sand, and then there appears to be a pause, the fine particles of clay all remain in suspension ; but finally, when the water is quite motionless, they begin to settle ; they fall very slowly, however, and the water does not become clear for hours.

We should, however, imitate the natural conditions much more closely by strewing the detritus in a current of water of varying velocity.   The gravel would then be dropped at points where the current was still very strong, and then as the velocity of the current diminished, the coarse and fine sands would be deposited in succession in different places, while the clay would remain in suspension until the water became perfectly quiet, as when a stream enters a lake or the sea, and then it would be slowly deposited.

These are very instructive experiments.   We learn from them :—

First, that the power of the water to hold particles in suspension is inversely proportional to the size of the particles.

Second, that all materials deposited in water are assorted according to size, as shown by the specimens of gravel, sand, and clay (**41**) from the shore near Boston.

Third, and this is one of the most important facts in geology, all water-deposited sediments are arranged in horizontal layers, *i.e.*, are stratified, as shown more clearly in the specimens of conglomerate, sandstone, and slate (**42**).

We have now traced to its conclusion, though very briefly, the process of the formation of one great division of *stratified* rocks, — the *mechanically-formed* or *fragmental* rocks. These are so-called because the clay sand, and gravel are, in every instance, fragments 'd pre-existing rocks; and because the formation, transportation, and especially the *deposition*, of these fragments, are the work chiefly or entirely of mechanical forces.

All the fragmental rocks are, at the time of their deposition, more or less carefully assorted gravel, sand, or clay; the coarser kinds, like gravel, containing, usually, only enough of the finer material — sand or clay — to fill the interstices between the fragments. After their deposition, the gravel, sand, and clay are usually slowly consolidated or hardened to form con·glomerate, sandstone, and slate; but the causes of the consolidation may be most conveniently explained in the guide to the Lithological Collection in Room B.

CHEMICAL DEPOSITION.

It is a well-known fact that the sea holds in solution vast amounts of common salt as well as many other substances; and, as already pointed out, analyses of river-waters show that dissolved minerals derived from the chemical decomposition of the rocks of the land are being constantly carried into the sea.

Portions of the sea which are cut off from the main body, and which are gradually drying up, like the Great Salt Lake and Dead Sea, become saturated solutions of the various dissolved minerals, and these are slowly deposited. This process is very clearly illustrated along our shores in summer, where, during storms, salt-water spray is thrown above the reach of the tides, and, collecting in hollows in the rocks, gradually dries up, leaving behind a crust of salt.

When water lays down matter which it held in *suspension*, we call the process *mechanical* deposition, and the result is *mechanically*-formed rocks. But when it lays down matter which it held in *solution*, we call the process *chemical* deposition, and the result is *chemically*-formed rocks.

The principal substances which water deposits chemically are common salt, forming beds of rock-salt (**43**) ; sulphate of calcium, forming beds of gypsum (**44**) ; carbonate of calcium, forming beds of limestone (**46-47**) ; and the double carbonate of calcium and magnesium, forming beds of dolomite (**45**). Inorganic deposition, like inorganic erosion, is thus both chemical and mechanical.

ANIMALS AND PLANTS, OR ORGANIC AGENCIES.

As regards the *destruction* of rocks — *erosion* — plants and animals are almost powerless; and a single illustration (**61**) will suffice. This is a fragment of limestone which has been perforated, and thus partly worn away, by a bivalve mollusk. In the role of *rock-makers*, on the other hand, organisms play a very important part, being very efficient agents of *deposition*.

FORMATION OF COALS AND BITUMENS.

The general physical conditions under which peat (**62**) is formed are familiar facts. We require simply low, level land, covered with a thin sheet of water and abundant vegetation ; in other words, a marsh or swamp. If plants decay on the dry land, the decomposition is complete ; they are burned up by the oxygen of the air to *carbon dioxide* and *water* just as surely as if they had been thrown into a furnace, though less rapidly, and nothing is returned to the soil but what had been taken from it by the plants during their growth. But if the plants decay under water, as in a peat-marsh or bog, the decay is incomplete, and most of the carbon of the wood is left behind. Now, if this incomplete combustion of vegetable tissues takes place in a charcoal-pit, where the wood is out of contact with air from being covered with earth, we call the carbonaceous product charcoal; but if under the water of a marsh, in nature's laboratory, we call the product peat. Peat is simply a natural charcoal ; and, just as in ordinary charcoal, its vegetable origin is always perfectly evident. But when the deposit becomes thicker, and especially when it is buried under thick formations of other rocks, like sand and clay, the great pressure consolidates the peat; it becomes gradually more mineralized and shining, shows the vegetable structure less distinctly, becomes more nearly pure carbon, and we call it in succession lignite (**63**), bituminous coal (**64**), and anthracite (**65**).

This is, briefly, the way in which all varieties of coal, and some of the more solid kinds of bitumen, like asphaltum, are formed. But the lighter forms of bitumen, such as petroleum and naphtha, appear to be derived mainly, if not entirely, from the partial decomposition of animal tissues. These, it is well-known, decay much more readily than vegetable tissues; and the water of an ordinary marsh or lake contains sufficient oxygen for their complete and rapid decomposition. In the deeper parts of the ocean, however, the conditions are very different, for recent researches have shown, contrary to the old idea, that the deep sea holds an abundant fauna.

All grades of animal life, from the highest to the lowest, have need of a constant supply of oxygen. On the land vegetation is constantly returning to the air the oxygen consumed by animals, but in the abysses of the ocean vegetable life is scarce or wanting; and hence it must result that over these greater than conti nental areas countless myriads of animals are living habitually on short rations of oxygen, and in water well charged with carbon dioxide, the product of animal respiration. As a conse-quence, when these animals die their tissues do not find the oxygen essential for their perfect decomposition, and in the course of time become buried, in a half-decayed state, in the ever-increasing sediments of the ocean-floor.

It is important to observe that an abundance of organic matter decaying under water is not the only condition essential to the formation of beds of coal and bitumen; for this condition is realized in the luxuriant growth of sea-weeds fringing the coast in every quarter of the globe; and yet coals and bitumens are rarely of sea-shore origin. These organic products, even under the most favorable circumstances, accumulate with extreme slowness; far more slowly, as a rule, than the ordinary mechanical sediments, like sand and clay, with which they are mixed, and in which they are often completely lost. Consequently, although the deposition of the carbonized remains of plants and animals is taking place in nearly all seas, lakes, and marshes, it is only in those places where there is little or no mechanical sediment that they can predominate so as to build up beds pure enough to be called coal or bitumen. In all other cases we get merely more or less carbonaceous sand or clay. Now these especially favorable localities will manifestly not be often found along the sea-shore, where we have strewn the sand and clay brought down by rivers or washed off the land directly by the ever-active surf; but they must exist in the central portions of the ocean, where there is almost no mechanical sediment and yet an abundance of life, and in swamps and marshes, where there is scarcely sufficient water to cover the vegetation, and no waves or currents to wash down the soil from the surrounding hills.

FORMATION OF IRON ORES.

The iron ores are another class of rocks which are formed through the agency of organic matter. All rocks and soils contain iron, but it is mainly in the form of the insoluble peroxide, and hence cannot be soaked out of the soil by the rain-water and concentrated by the evaporation of the water at lower levels in ponds and marshes, as a soluble substance like salt would be. If carried off with the sand and clay, by the mechanical action of water, it remains uniformly mixed with them, and there is but little tendency to its separation and concentration so as to form a true ore of iron.

But what water cannot do alone is accomplished very readily when the water is aided by decaying organic matter, which is always hungry for oxygen, being, in the language of the chemist, a powerful reducing agent. The soil, in most places, has a superficial stratum of vegetable mould or half-decayed vegetation. The rain-water percolates through this and dissolves more or less of the organic matter, which is thus carried down into the sand and clay beneath and brought in contact with the ferric oxide, from which it takes a certain proportion of oxygen, reducing the ferric to the ferrous oxide. At the same time the vegetation is burned up by the oxygen thus obtained, forming carbon dioxide, which imme-diately combines with the ferrous oxide, forming carbon-ate of iron, which, being soluble under these conditions, is carried along by the water as it gradually finds its way by subterranean drainage to the bottom of the valley and emerges in a swamp or marsh.

Here one or two things will happen : if the marsh con-tains but little decaying vegetation, then as soon as the ferrous carbonate brought down from the hills is exposed to the air it is decomposed, the carbon dioxide escapes, and the iron, taking on oxygen from the air, returns to its original ferric condition ; and being then quite insol-uble, it is deposited as a loose, porous, earthy mass, commonly known as bog-iron ore (**66**), which becomes gradually more solid and finally even crystalline through the subsequent action of heat and pressure. When first deposited, the ferric oxide is combined with water or hydrated, and is then known as limonite ; at a later period the water is expelled, and we call the ore hema-tite (**67**) ; and at a still later age it loses part of its oxygen, becomes magnetic and more crystalline and is then known as magnetite (**68**). Thus it is seen that the iron ores, as we pass from bog limonite to magnetite, form a natural series similar to and parallel with that afforded by the coals as we pass from peat to anthracite. If the drainage from the hills is into a marsh containing an abundance of decaying vegetation, *i.e.*, if peat is forming there, the ferrous carbonate, in the presence of the more greedy organic matter, will be unable to obtain oxygen from the air ; and as the evaporation of the water goes on, it will sooner or later become saturated with this salt, and the latter will be deposited **(69)**. Here we find an explanation of a fact often observed by geologists, viz., that the carbonate iron ores are usually associated with beds of coal.

The formation of the iron ores, like that of the coals and bitumens, is a slow process; and the ores, like the coals, etc., will be pure only where there is a complete absence of mechanical sediment, a condition that is realized most nearly in marshes.

FORMATION OF LIMESTONE, DIATOMACEOUS EARTH, ETC.

Aquatic organisms take from the water certain mineral substances, especially silica and carbonate of calcium, to form their skeletons. Silica is used only by the lowest organisms, such as Radiolaria, Sponges, and the minute unicellular plants, Diatoms. The principal animals secreting carbonate of calcium are Corals and Mollusks. These hard parts of the organisms remain undissolved after death ; and over areas where there is but little of other kinds of sediment they form the main part of the deposits, and in the course of ages build up very extensive formations which we call diatomaceous earth or tripolite if the organisms are siliceous, or limestone if they are calcareous.

The specimen of diatomaceous earth or tripolite (83) represents that which is now forming in ponds and marshes near Boston ; and deep-sea dredging has proved that a similar impalpable siliceous ooze or earth is now accumulating over very extensive areas of the ocean-floor.

The next two specimens (82, 84) show very clearly how limestones are formed by the accumulation of shells. At first the shells are loose or unconsolidated (82) ; but they are soon firmly cemented together by the deposition of carbonate of calcium between them (84). This newly formed limestone or coquina is a very porous, friable rock ; but in the course of time the interstices between the shells become filled with finer fragments and clay, and the coquina is gradually changed to the dense, solid, fossiliferous limestone (85) found in the older formations.

Limestones are also made in a similar manner from corals (86). The pure coralline limestones, however, are formed chiefly on coral islands and reefs, since the rock-building corals, unlike mollusks, do not flourish in all parts of the sea. They are, in general, limited to regions where the temperature of the sea-water does not fall below 68° Fahr. and the depth does not exceed 120 feet, i.e., to shallow water in the tropics or warmer regions of the globe.

Coral polyps extract carbonate of calcium from sea-water and deposit it within their own bodies. The radiated structure of each polyp is perfectly reproduced in its coralline axis. This is a purely vital function, having no more connection with volition than the secretion of the shell of an oyster or the bones of the higher animals. The carbonate of lime thus deposited within the animal constitutes 90 to 95 per cent. of its whole weight. A single coral polyp is very small, but, like many of the lower animals, it has the power of multiplying indefinitely by buds and branches. Thus are formed compound corals (87). These may branch profusely, forming coral trees, or grow in hemispherical masses, called coral heads.

Coral polyps also reproduce by eggs; and thus from one coral tree other coral trees spring up all around and form a coral forest. Finally, the limestone accumulation of thousands of generations of the coral-forest growing and dying on the same spot, together with the shells of mollusks and the bones of fishes, the whole, of course, crowned with the living forest of the present generation, constitute the coral reef. It is evident, then, that a reef is formed somewhat after the manner of a peat bog. As a peat bog repres.. ′   ′ matter taken from the air, so a coral reef repre:.   . matter taken from the sea-water. As each genera⸗   self to the ancestral funeral pile, the reef steadily rises until it becomes elevated far above the surrounding sea-bottom.

The models (88–89) illustrate the principal phases of the growth of coral reefs around oceanic islands. The first model shows fringing reefs only. The land is in a state of rest, as regards movements of elevation and subsidence, and the reef-building corals have formed a fringe around the islands, extending out as far as the depth will allow and building nearly straight up from that line to the surface of the water. The fringing reef, represented by the white band following the shore-lines of the islands, is essentially a coral terrace or platform with a breadth inversely proportional to the angle of slope on which it is built, and terminating outwards in a steep, straight wall of coral extending down to the maximum dept'ı of twenty fathoms. When the fringing reef reaches the surface of the water, its ı growth must stop, unless the water is made deeper by a gradual subsidence of the sea-bottom. This gradual subsidence has occurred in the case of nearly all extensive reefs. As the original island slowly sinks beneath the waves, the outer, most rapidly growing, edge of the reef keeps pace with the downward movement, leaving a shallow, circular channel of water between it and the shore, as shown in the second model. This outer ring of coral is called a barrier reef; and the channel inside of it owes its existence to the fact that the coral grows most rapidly on the outer edge of the reef, where the water is freshest and purest.

The piece of phosphate rock (**81**) represents, in a general way, the accumulation of the bones and excrement of the higher animals, in which phosphate of lime is the chief mineral constituent.

The rocks here considered may be, and, as we have already seen, sometimes are, deposited in a purely chemical way, without the aid of life; and it is important to observe that in no case do the organisms make the silica and carbonate of calcium of their skeletons, but they simply appropriate and reduce to the solid state what exists ready-made in solution in the sea-water. These minerals, and others, as we know, are produced by the decomposition of the rocks of the land, and are being constantly carried into the sea by rivers; and, if there were no animals in the sea, these processes would still go on until the sea-water became saturated with these substances, when their precipitation as limestone, etc., would necessarily follow. Hence it is clear that the animals simply effect the precipitation of certain minerals somewhat sooner than it would otherwise occur; so that from a geological standpoint the differences between chemical and organic deposition are not great.

This section of our subject may be summarized as follows: Animals and plants contribute to the formation of rocks in three distinct ways:—

1. During their growth they deoxidize carbon dioxide and water, and reduce to the solid state in their tissues, carbon and the permanent gases oxygen, hydrogen, and nitrogen ; and after death, through the accumulation of the half-decayed tissues in favorable localities, — marshes, etc., — these elements are added to the solid crust of the earth in the forms of coal and bitumen.

2. During the decomposition, i. e., oxidation, of the organic tissues, the iron existing everywhere in the soil is partially deoxidized, and, being thus rendered soluble, is removed by rain-water and concentrated in low places, forming beds of iron ore.

3. Through the agency of aquatic organisms, certain mineral substances are being constantly removed from the water and deposited upon the ocean-floor, forming various calcareous and siliceous rocks.

We now bring the review of the aqueous or superficial agencies to a conclusion by noting once more that the great geological results accomplished by *air*, *water*, and *organic matter* or *life* are: (1) *Erosion*, or the wearing away of the surface of the land ; and (2) *Deposition*, or the formation from the debris of the eroded land of two great classes of stratified rocks, — the mechanically-formed or fragmental rocks, and the chemically and organically-formed rocks.

STRUCTURAL GEOLOGY.

Structural Geology treats of the different kinds of minerals and rocks, and rock-structures; and, in its broadest aspect, takes account of the constitution and contours of the entire crust of the earth. This department of geology embraces two distinct sciences: Mineralogy, which treats of the composition, structure, and, physical properties of homogeneous chemical compounds or minerals: and Petrography, which treats of the composition, structure, and distribution of rocks, or the massive, impure aggregates of minerals.

The mineralogical collections are in Room A and the Petrographical collections in Room B. The Guide to Mineralogy has been published as a separate volume, and this volume embraces the second division, only, of Structural Geology, — Petrography.

Petrography, the science of rocks, is conveniently divided into two subordinate sciences, Lithology and Petrology. Lithology is an in-door science; we use the microscope largely, and work with hand specimens or thin sections of the rocks, observing the composition and those small structural features which go under the general name of texture. In Petrology, on the other hand, we consider the larger kinds of rock-structure, such as stratification, jointing, folds, faults, cleavage, etc.; and it is essentially an out-door science, since to study it to the best advantage we must have, not hand specimens, but ledges, cliffs, railway-cuttings, gorges, and mountains. The lithological collections occupy the wall-cases (sections 1—23) in Room B and the petrological collections the two large central or floor-cases.

LITHOLOGY.

A *rock* is any mineral or mixture of minerals occurring in masses of considerable size. This distinction of size is almost the only one that can be made between rocks and minerals, and that is very indefinite. A rock, however, whether composed of one mineral or several, is always an aggregate; and therefore no single crystal or mineral-grain can be properly called a rock.

Still, the study of minerals is not the same, by any means, as the study of rocks. In mineralogy each mineral species is regarded as a definite chemical compound—the composition being expressible by a chemical formula—and the aim is to study each mineral separately and to observe its chemical and physical characters in the purest state in which it occurs in nature. The mineralogist likes to find minerals separate and pure, and seeks to avoid mixtures.

In lithology, on the other hand, we deal mainly with mixtures of minerals, and the notion of definite composition has to be almost entirely given up; for there are only a very few minerals, such as quartz and calcite, which by themselves form masses large enough to be called rocks (*simple rocks*, such as limestone, quartzite, etc.), most rocks being composed of several minerals more or less intimately mixed in small particles (*mixed rocks*, such as granite, mica schist, etc.). We seldom find anything quite pure in lithology, but nature mixes the minerals in ever varying proportions.

Since rocks are mixtures or aggregates of minerals, it follows that they must be usually more complex compounds than minerals; for the various chemical elements are first combined to make minerals, and then the minerals are combined to form rocks. The combination of the elements in minerals is chemical and definite; while the combination of the minerals in rocks is mechanical and indefinite.

Rocks are composed of solid matter, but the particles are not necessarily consolidated or cemented together. Thus, the loose sand and gravel on the beach are as truly rocks, in the geological sense, as sandstone and conglomerate. Clay is a rock as much as slate or granite.

Lithology, like mineralogy, is either comparative or systematic, according as we make the properties of rocks, or the individual kinds or species, families, and classes of rocks, the special subject of investigation; and in a complete educational arrangement *Comparative Lithology* properly precedes *Systematic Lithology*.

COMPARATIVE LITHOLOGY.

The natural characteristics or properties of rocks are nearly all comprised under the two general heads of *composition* and *texture*.

COMPOSITION OF ROCKS.

Rocks are properly defined as large masses or aggregates of mineral matter, consisting in some cases of one and in other cases of several mineral species. Hence it is clear that the composition of rocks is of two kinds: chemical and mineralogical.

Chemical Composition of Rocks.

The elementary substances of which rocks are chiefly composed, which make up the main mass of the earth so far as we are acquainted with it, number only fourteen :—

Non-Metallic or Acidic Elements.— Oxygen, silicon, carbon, sulphur, chlorine, phosphorus, and fluorine.

Metallic or Basic Elements.— Aluminum, magnesium, calcium, iron, sodium, potassium, and hydrogen.

The elements are named in each group in about the order of their relative abundance ; and to give some idea of the enormous differences in this respect it may be stated that two of the elements — oxygen and silicon — form more than half of the earth's crust.

It is also noteworthy that, with the exception of iron, all the metallic elements named are light and light-colored. Iron is the great coloring agent in rocks, and all the heavier rocks contain a large proportion of iron. Examples of all these rock-forming elements will be found in the complete series of the chemical elements in the first or north floor-case in the mineralogical room (A).

Mineralogical Composition of Rocks.

The fourteen elements named above are combined to form about fifty minerals with which the student of lithology should be acquainted, although not more

[1] These two sections are to be regarded as one, although each has a separate series of numbers. The specimens on each shelf are continuous from section 1 to section 2.

than one half of these are of the first importance. Typical specimens of the principal species are arranged in sections 1 and 2. These rock-forming minerals may, for lithological purposes, be classified in several ways :

1. Lithologists, when considering chiefly the definitions of rocks, distinguish their constituents as *essential* and *accessory*.

The essential constituents of a rock are those minerals which are essential to the definition of the rock. For example, we cannot properly define granite without naming quartz and orthoclase; hence these are essential constituents of granite; and if either of these minerals were removed from granite it would not be granite any longer, but some other rock. But other minerals, like tourmaline and garnet, may be indifferently present or absent; it is granite still. Hence they are merely accidental or accessory constituents. They determine the different *varieties* of granite, while the essential minerals make the *species*. This classification, of course, is not absolute, for in many cases the same mineral forms an essential constituent of one rock and an accessory constituent of another. Thus, quartz is essential in granite, but accessory in diorite.

2. When considering chiefly the origin and subsequent history of rocks, it is more important to distinguish their constituents as *original* and *secondary*.

The original constituents of a rock are those minerals of which it was composed when first formed; while the secondary constituents are those species which have resulted from the decomposition or alteration of the original constituents. With the eruptive rocks, especially, this distinction is often an important one, since the present composition is, with the older eruptives at least, often widely different from the original composition. This classification, like the preceding, holds only in a general way, the same mineral, in many cases, occurring as an original constituent of certain rocks and as a secondary constituent of other or even of the same rocks. It would manifestly be impossible to fully illustrate these classifications without duplicating the series of minerals; the labels, however, state distinctly for each species whether, in its usual mode of occurrence, it is chiefly essential or accessory, and whether it is chiefly original or secondary. These distinctions will also receive due attention in the following descriptions of the rock-forming minerals; and in the systematic collection of rocks both the original and secondary constituents of each group of rocks, as well as the essential and accessory constituents, will be fully illustrated.

The actual arrangement of the minerals is essentially mineralogical, but with special reference to their occurrence in rock masses, the object being to keep together species that are naturally associated in the rocks and are hence closely related as rock constituents.

Besides the chemical composition of the minerals, which it is impracticable to illustrate in a satisfactory way, and the more obvious morphologic and physical properties, such as form, color, luster, etc., which may be readily appreciated in the specimens themselves, and are fully explained in the mineralogical collection and guide, there are in most of the species important optical or structural features which are essentially microscopic. In fact these features, although throwing much light upon the origin and history of rocks, usually require polarized light as well as the compound microscope for their exhibition. The only resource, therefore, is either to wholly omit them, which their importance forbids, or to illustrate them as far as practicable pictorially. This is the object of the diagrams and colored drawings interspersed through the collection.

SILICA AND SILICATES. — This is by far the most important class of rock-forming minerals. It includes

free silicic acid or silica in the forms of quartz and opal,
and the principal silicates, or the minerals formed by
the union of silica with various metals. The silicates
embrace more than one fourth of the known species of
minerals and, omitting quartz and calcite, all of the really
important constituents of rocks. They may be naturally
divided into two great groups, the *basic* and *acidic.*

These groups are not sharply defined, on the contrary there
is a perfectly gradual passage from one to the other; and yet
this is, for geological purposes at least, a very natural classification. The dividing line falls in the neighborhood of 60 per
cent. of silica: *i. e.*, all species containing this proportion of
silica or less are classed as basic, since in them the basic elements predominate; while those containing more than 60 per
cent. of silica are classed as acidic, because their characteristics are determined chiefly by the acid element or silica. The
principal bases occurring in the silicates, named in the order of
their relative importance, are aluminum, magnesium, calcium,
iron, sodium, and potassium; and of these, magnesium, calcium, iron, and usually sodium, are especially characteristic
of basic species. Iron is the heaviest base and consequently
the basic must be, as a rule, heavier as well as darker colored
than the acidic silicates. All this is of especial importance
because in the rocks nature keeps the basic and acidic silicates
separate in a great degree.

A more convenient division, and the one chiefly observed in the arrangement of the specimens, is that distinguishing the anhydrous from the hydrous silicates.
These two classes are rarely mingled in the rocks; and
this distinction is only second in importance to, and
more practicable than, that between the basic and
acidic silicates.

Silica.— This extraordinarily abundant and protean substance occurs in crystalline forms, as quartz, tridymite, etc., and amorphous, as opal.

Quartz.— The varieties of quartz are very numerous, but they are naturally divided into two groups, as follows, only the most abundant, rock-forming varieties in each group being included in this collection :—

1. *Phenocrystalline* or *vitreous* varieties, including the distinctly crystallized quartz or rock-crystal (**1**), vitreous quartz (**6**), granular quartz (**2**), milky quartz (**7**), etc. These varieties best represent quartz in its occurrence as an original and essential constituent of such abundant rocks as granite, mica schist, sandstone, etc.

2. *Cryptocrystalline* or *compact* varieties, including chalcedony (**4**), agate (**8**), flint (**11**), chert (**12**), etc. These varieties are, to a large extent, secondary and accessory constituents of the rocks in which they occur.

Phenocrystalline quartz occurs chiefly in hexagonal crystals, or in irregular vitreous masses or grains, devoid of cleavage; while the cryptocrystalline varieties are commonly characterized by botryoidal, concretionary, geoditic, and banded forms, all of which as well as the physical properties are fully explained in the mineralogical guide.

The figures (**3**), represent on a highly magnified scale, the numerous liquid inclusions which are the most important and constant microscopic feature of crystalline, rock-forming quartz. The right hand figure shows the usual arrangement of the inclusions in rows or bands; while the other figure, which is more highly magnified, shows the forms of the individual inclusions more accurately and the inclosed bubbles and crystals.

The enclosed liquids are usually either liquid carbon dioxide, water, or aqueous solutions, the nature of the dissolved salt being often indicated by a small crystal of it which separates from the solution under diminished temperature or pressure. The liquid usually fails to fill the cavity, the vacant space appearing like the bubble in a spirit-level and indicating that the quartz probably crystallized at a high temperature, the liquid which became enclosed in the growing crystal shrinking on subsequent cooling.

Quartz is, next to common feldspar, the most important constituent of the earth's crust; and it is also the hardest and most durable of all common minerals. It is not decomposed by the action of the elements; and, being very hard and devoid of cleavage, the same mechanical wear which, assisted by more or less chemical decomposition, reduces softer minerals to an impalpable powder or clay, leaves quartz chiefly in the form of sand and gravel.

Opal.—This species is like quartz in composition, except that it contains from 3 to 20 per cent. of water. It is absolutely devoid of crystalline structure, but presents a series of varieties similar to those of cryptocrystalline quartz. It is softer, lighter, and more soluble than quartz; and usually shows a tendency to slowly lose its water and change into quartz. The more beautiful and interesting varieties, such as precious opal, are very rare and of little geological importance. The semi-opal (5) resembles jasper and chalcedony and is found under similar conditions. But the specially important geological varieties of opal are *geyserite* or *siliceous tufa* (9), which is hydrated silica or opal deposited by hot springs and geysers; and *tripolite* or *diatomaceous earth* (10), which is opal in the form of the microscopic siliceous shells of some of the lowest types of animals and plants.

Anhydrous Silicates. — This division of the silicates is made up, so far as the rock-forming kinds are concerned, of four important and natural groups of species which occur chiefly as essential constituents of rocks: *feldspars, feldspathides, micas,* and *hornblende* and *augite;* and one large group of accessory constituents. These groups are named here in approximately their order of importance or abundance; and in this order they exhibit a gradation in the proportions of alkalies (potash and soda) which the species contain, the feldspars being richest in alkalies, while the accessory species contain little or no alkalies.

Feldspars. — Feldspar is the name, not of a single species, but of a family of minerals. There are six principal feldspars, all of which are important constituents of rocks. The enormous abundance of the feldspars is expressed in the name, which is the German for *field-spar.* The name is sometimes spelled *felspar,* which is the German for *rock-spar,* implying that these are the common spars or minerals of the rocks and fields. The feldspars, with the kaolin or clay which results from their decomposition, undoubtedly form more than one half of the entire crust of the earth.

The feldspars may be classified chemically as follows :

Orthoclase (2-8), silicate of aluminum and potassium, or potash feldspar.

Albite (9-10), " " " " sodium, or soda feldspar.

Anorthite (15), " " " " calcium, or lime feldspar.

Oligoclase (12), " " " sodium and calcium, or soda-lime feldspar.

Andesite (13), " " " and equal parts of sodium and calcium.

Labradorite (14, 16), " " " calcium and sodium, or lime-soda feldspar.

This complex arrangement can be greatly simplified for geological purposes. Orthoclase crystallizes in the monoclinic system and all the other feldspars in the triclinic system. With the exception of albite, which is a comparatively rare species, the triclinic feldspars all contain less silica than orthoclase, *i. e*, are more basic. This is shown by the subjoined table giving the average composition of each of the feldspars.

	SiO_2	Al_2O_3	K_2O	Na_2O	CaO	Total.
Orthoclase,	65	18	17	—	—	100
Albite,	68	20	—	12	—	100
Oligoclase,	62	24	—	9	5	100
Andesite,	58	27	—	7	8	100
Labradorite,	53	30	—	4	13	100
Anorthite,	43	37	—	—	20	100

The triclinic or basic feldspars are commonly associated with each other and with other basic minerals, but are rarely important constituents of rocks containing much orthoclase. In other words, the distinction of orthoclase from the basic feldspars is important and comparatively easy; while the distinction of the different basic feldspars from each other is both unimportant and difficult. Hence, in lithology it is convenient and natural to class all the basic feldspars together, as if they were one species, under the name *Plagioclase*, which refers to the oblique cleavage of these feldspars, and contrasts with *Orthoclase*, which refers to the right-angled cleavage of that species.

Orthoclase, or the common feldspar, is probably equal in abundance to *plagioclase*, *i. e.*, to all the basic feldspars, taken together. Its principal rock-forming varieties are well shown in the specimens (**2–8**). The crystals (**2, 4**) are usually simple monoclinic forms; but the double crystals or Carlsbad twins (**3**) are almost equally characteristic, and afford a ready and certain means of distinguishing orthoclase from plagioclase. The rectangular cleavage of orthoclase (**5**) is, perhaps, its most important feature. The specimen of trachyte (**6**) shows large Carlsbad twins of *sanidin*, a clear variety of orthoclase which is a very important constituent of volcanic rocks. The compact form of orthoclase is best represented by felsite (**7**), and the glassy or perfectly amorphous form by obsidian (**8**). Felsite and obsidian are volcanic rocks chiefly composed of orthoclase, which have solidified too suddenly to permit distinct crystallization. The most characteristic microscopic features of orthoclase are represented in the figures (**1**). The one on the left hand shows a peculiar striping supposed to be due to the interlamination of orthoclase and albite, and the other a curious cross-hatching or reticulation with brilliant colors, often observed in orthoclase in polarized light.

Albite, although triclinic in crystallization, is the most acidic of all the feldspars; and it is so generally associated with orthoclase in the rocks, that it may be fairly classed with that species. The specimens represent both the coarsely ciystalline (9) and the massive (10) forms of albite, the former showing the parallel lines or twinning striae which indicate, without farther examination, that it is a triclinic species.

Plagioclase is represented in the collection by all of its component species, but the two specimens of *Labradorite* (**14, 16**) are decidedly the most typical examples. They show that plagioclase is contrasted with orthoclase and albite by its darker color. The coarsely crystalline or cleavable specimen (**14**) shows the beautiful play of colors, which is sometimes observed in other feldspars, but is especially characteristic of this species, and the straight, parallel lines or bands due to repeated twinning. These twinning striae are exceedingly characteristic of plagioclase; and, although often so fine as to be invisible to the unassisted eye, they are strongly contrasted under the microscope in polarized light, and afford the best means of distinguishing orthoclase and plagioclase.

The figure (**11**) represents typical sections of plagioclase magnified in polarized light. The extremely important fact that plagioclase is decomposed or kaolinized when exposed to the weather much more rapidly than orthoclase is illustrated by the weathered specimens of granite (**17**) and diabase (**18**), from this vicinity, for orthoclase is the principal constituent of the former rock and plagioclase of the latter. These specimens have probably been exposed to the weather for equal times; and yet the granite shows but little change, while the diabase, which was also once a firm, hard rock has been thoroughly rotted and reduced to a fine soil.

Feldspathides. — The minerals of this group are called feldspathides for the reason that, while strikingly

unlike the true feldspars in crystallization, they are
almost identical with them in composition and physical
characters, and, furthermore, they commonly appear
as substitutes for the feldspars in the rocks. They are
far less abundant than the feldspars, and are, on the
average, more basic in composition and association, occurring chiefly in the more basic eruptive rocks. Only
the more abundant, rock-forming feldspathides are represented in the collection (**21-26**).

Nephelite, which is similar to albite in composition,
except that it contains much less silica, crystallizes in
hexagonal prisms, and presents two distinct varieties.
The typical nephelite (**25**) forms small glassy crystals in
volcanic rocks ; while the variety *elaeolite* (**23**) is coarsely
cystalline or cleavable, translucent, has a greasy luster
and is found chiefly in the older plutonic rocks. Typical nephelite stands in the same relation to elaeolite that
sanidin does to orthoclase.

Sodalite (**26**) and *Hauynite* (**24**), the latter being represented by the single blue crystal in the dark colored
lava, are rarer, isometric species occurring under much
the same conditions as nephelite.

, *Leucite* (**21**), which contains the same elements as orthoclase, but with only 55 per cent. of silica, is especially
interesting on account of its very perfect isometric form.

The crystals are tetragonal trisoctahedrons; but the shaded
figure (**22**), which represents the appearance of a thin section in
polarized light, shows in the double refraction that the optical
characters are inconsistent with this form, since all crystals in
this system should, theoretically, be strictly singly-refracting.
When the leucite cryrtals are heated, however, the isometric

symmetry is developed internally as well as externally, indicating that the abnormal optical characters are the consequence of strains due to cooling from the high temperatures at which the crystals were formed. The other figure shows the concentric arrangement of the enclosed crystallites.

Micas. Mica, like feldspar, is not the name of a single mineral, but of a group or family of minerals, including some half-dozen species **(27–32)**. Four of these are represented in the collection, although only two — muscovite and biotite—are of the first importance as constituents of rocks. All the micas are basic silicates of alumina and potash; and certain kinds are also rich in magnesia and iron. The crystallization is monoclinic; but all the species occur commonly in flat rhombic or hexagonal forms **(27)**. Undoubtedly the most important and striking characteristic of the whole mica family is the remarkably perfect cleavage parallel with the basal planes of the crystals **(28–29)**, and the wonderful thinness and elasticity of the cleavage lamellae. The cleavage contrasts the micas with all other common minerals, and makes their certain identification one of the easiest things in lithology.

Muscovite is distinctly more acidic than *biotite*, both in composition and in associated minerals. It contains little or no iron and magnesium as against 20 per cent. of oxide of iron and 17 per cent. of magnesia in biotite. The large proportion of iron in biotite fully explains the difference in color, which is usually so marked that muscovite is known as the *white* mica and biotite as the *black* mica.

The crystal of muscovite **(27)** represents equally well the form of biotite. The large plates **(28–29)**, as well as the distinct crystals of mica, are found only in veins, the mica in common rocks, such as granite, gneiss, mica schist (30), etc., occurring for the most part in thin scales of indefinite or irregular outlines. Muscovite is the mica of commerce, being colorless and clear in thin plates.

Phlogopite (3I), and *lepidomelane* (32), are much rarer species, although they are occasionally important rock-forming minerals. Phlogopite is a highly magnesian mica, usually of a brownish color, and is found chiefly in crystalline magnesian limestone, dolomite, serpentine, and other magnesian rocks. Lepidomelane is only slightly magnesian, but it is distinctly a ferruginous mica, containing about 30 per cent. of iron oxide. It resembles biotite, and occurs in various granitic and metamorphic rocks.

Hornblende and Augite.—These two minerals are the common, rock-forming varieties of the species *Amphibole* and *Pyroxene*, which may be described as very similar complex silicates, similar in composition, crystallization and physical characters, and in their numerous varieties depending on variations in composition and structure. These varieties include, besides hornblende and augite, which are black, the green varieties, actinolite, sahlite, etc., and the white varieties, tremolite, diopside and as- bestus. All these varieties are of interest to the mineralogist, but *hornblende*, the black, aluminous, and highly ferruginous variety of amphibole, and *augite*, the corresponding variety of pyroxene, are the only ones that are sufficiently abundant to be of special geological interest. Hence the geologist naturally substitutes the names of

.

these two important varieties for the names of the species
to which they belong.

Hornblende and *Augite* rank in abundance with the
black and white micas. But they are not so easily distinguished as the micas, for, as already indicated, they
are essentially identical in composition and physical
characters, and they farther agree in crystallizing in the
monoclinic system.

It might appear at first that the distinction of minerals so
nearly identical is not an important matter. But nature has de-
creed otherwise, for, as with the micas and feldspars, they are
to a large extent kept separate in the rocks. In associated minerals, augite is distinctly more basic than hornblende. In proof
of this we need only to know that it rarely occurs in the same
rock with original quartz, while hornblende is found very commonly in that association. It is these differences of association,
chiefly, that make the distinction of hornblende and augite essential to the proper recognition of rocks. The specimens
have, therefore, been selected with special reference to showing how the two minerals are contrasted and may be distinguished.

The crystallized specimens are intended especially to
show the angles of the monoclinic prisms, the prismatic angle which is 124° 30', or distinctly oblique, in hornblende
(**21**), is only 87° 5', or nearly a right angle, in augite
(**28**). In each mineral the principal cleavage is prismatic,
and hence the cleavage fragments (**26-29**) yield the
same angles. In both the crystals and the cleavage specimens the particular edges or angles referred to are designated by the small strips of paper which are bent over
them. The importance of the cleavage angles in the

microscopic study of these minerals may be readily in-
ferred from the drawings (22). The first figure, showing
cleavage lines crossing obliquely, represents hornblende ;
and the second one, showing cleavage lines crossing
nearly at right angles, represents augite. The massive
specimen of hornblende (31) might be readily duplicated for augite ; but not so with the specimen of bladed
hornblende (27), since this slender, bladed crystallization, although very characteristic of hornblende, is rarely
seen in augite. The principal optical feature distinguishing hornblende and augite is the dichroism (23),
which is very strong in hornblende (the green and yellow
crystals in the drawing) and wanting in augite.

The remaining specimen of hornblende (32) represents the
very common pseudomorphs of hornblende after augite, or the
variety *uralite*. The crystals of uralite are strictly *paramorphs*,
since the conversion of augite to hornblende is a change of
molecular structure only, and not of composition. Although the
molecular structure of augite appears to be stable with the conditions under which the mineral usually crystallizes in eruptive
rocks, it is relatively unstable under the existing conditions, in
many cases, since we find that, while the crystals retain the external forms of augite, they gradually take on the molecular
arrangement or cleavage of hornblende; and hornblende derived in this way is known as *uralite*. The next drawing (24),
shows crystals of augite which have been partially changed to
hornblende, the inner portion still showing the right-angled
cleavage of augite.

Enstatite (25) and *Hypersthene* (30) are closely related
to augite and hornblende both mineralogically and geologically, occurring very abundantly in some of the basic
eruptive rocks with augite or as a substitute for augite.

They crystallize in the orthorhombic system, but with essentially the same prismatic angles as augite. This is seen most clearly in the specimen of enstatite, which is a nearly perfect crystal. The fibrous appearance and metalloidal luster of the specimen of hypersthene are also very characteristic. Chemically these species are silicates of magnesia and iron, differing from the preceding group chiefly in the absence of lime.

Accessory Silicates. — The species of this group, as already noted, are generally non-alkaline ; and they are farther characterized, as a rule, by very distinct crystallization, and by great hardness. They may be conveniently divided, according to their modes of occurrence, into two sub-groups, the first three species in the following list being found chiefly in eruptive rocks and the remainder chiefly in the metamorphic sedimentary rocks, and especially in the schists. With the exception of epidote, they are all to be regarded as usually original constituents of the rocks in which they occur.

Chrysolite or *Olivine* (**41, 51**) is a very basic silicate of magnesium and iron, occurring chiefly in basalt and other highly basic eruptive rocks. It is a hard, green, glassy mineral, usually forming small, isolated crystals or grains in the dark-colored lavas.

The first specimen (**41**) shows grains of olivine that have been set free by the disintegration of the enclosing rock. Occasionally the grains are aggregated to form lumps or nodules in the lava, and sometimes they are the principal or even the sole constituent of the rocks, as in peridotite and dunite (**51**). Chrysolite changes very readily, by simple hydration, into serpentine; and it is the general belief of geologists that much of the serpentine in the rocks has had this origin.

Zircon (**46**) is a silicate of the rare metal zirconium (more probably, perhaps, a double oxide of silicon and zirconium), occurring quite commonly, but often microscopically, in syenite and allied rocks. The crystals, as the specimens plainly show, are simple tetragonal prisms and pyramids, of adamantine luster and great hardness.

Epidote (**42, 47**) is one of the most important of the secondary minerals in basic rocks, occurring usually as an alteration-product of augite, hornblende, and allied species. It is found chiefly in irregular veins and cavities in the rocks, in both distinctly crystallized (**42**) and massive (**47**) forms. It is a hard mineral and the yellowish green color is very characteristic.

Garnet (**49, 52**), in its numerous varieties, is the most abundant accessory in the schists and gneisses or metamorphic rocks, and is often found in granite and other eruptive kinds, sometimes as an essential constituent. Garnet varies widely in composition, but is exceedingly homogeneous in crystallization, occurring almost always in dodecahedrons and trisoctahedrons of the first system, so that it is readily recognized by its form. It is hard and heavy, and the color is usually reddish or brown.

Tourmaline (**44, 53**) is a highly complex and variable silicate. Its chief interest, chemically, lies in the fact that it contains from 4 to 10 per cent. of oxide of boron. It is hexagonal and usually hemihedral; occurring commonly in slender prisms having an approximately triangular cross section. It is hard, commonly black, strongly double refracting and dichroic. It is found for the most part in the same rocks as garnet, and especially in eruptive and vein granite and in quartz (**53**).

Andalusite (**43, 48**) is an important accessory of metamorphic states and schists.   It is as simple in composition as tourmaline is complex, and its usual form is the
simple orthorhombic prism.   The most abundant and interesting variety is *chiastolite* (**48**).   It contains a large
proportion of dark impurities, and these are distributed
in such a way as to give the appearance of a cross or
square on the section, as shown in the figures (**43**).

Cyanite (**54**) and *Fibrolite* (**55**) are the same chemical
substance as andalusite, occurring under similar conditions, but crystallizing in very slender triclinic and monoclinic forms, cyanite being especially characterized by
its flattened or bladed crystallization.

Staurolite (**45**) resembles the preceding species in
abundance and mode of occurrence.   It crystallizes in
simple orthorhombic forms ; but undoubtedly its leading
and specially distinguishing feature is the cruciform
twinning, which is also well represented by the specimens.

Chondrodite (**50**) is an important accessory of the older crystalline limestones.   Distinct crystals are uncommon, but it usually forms irregular grains of a yellowish or brownish color.

Hydrous Silicates. — All the rock-forming hydrous silicates belong to the section of Margarophyllites.   These
are, as a rule, imperfectly crystalline, foliated or compact, and soft, with, commonly, a greasy feel and greenish color.   They are, with unimportant exceptions,
original and essential (often the only essential) constituents of the various schists and other rocks in which they

occur. Their geological relations are well expressed by the following grouping :

Talcose or Magnesian Section. — Magnesia is the principal or only base in these species. They are usually soft, greenish and highly infusible ; and are common alteration products of the augite, hornblende, chrysolite, and other original constituents of basic eruptive rocks.

Talc (**41, 46**) is a very typical member of this section, being soft, greasy and usually greenish and foliated (**41**), although sometimes white and massive (**46**). It is found chiefly in an impure form in steatite or soapstone ; and the clear, foliated talc occurs principally in irregular veins traversing this rock.

Serpentine (**42**) is a protean mineral, occurring in many forms and colors ; but this compact, green variety is the one of greatest geological importance. In composition it is like talc with more water and twenty per cent. less of silica. Although it must be regarded as sometimes original and essential, yet it is, as stated, a common secondary constituent of basic eruptive rocks, and a very important accessary constituent of crystalline limestone and dolomite.

Chlorite (**47**) is a very basic species and contains alumina and iron as well as magnesia. Like serpentine, it is a very common secondary mineral ; but in chlorite schist we must regard the mineral, whatever its origin, as original and essential with reference to that particular species of rock. Chlorite is especially characterized by its dark green color and finely foliated structure.

Argillaceous or Aluminous Section. — Kaolin or clay is the chief constituent of this section, although it includes several less abundant species which agree with kaolin in containing little or no magnesia and in having been derived directly or indirectly from the feldspars and allied minerals.

Kaolinite or *Kaolin* (**43, 48**), the most abundant of all the hydrous silicates, is similar to serpentine in composition, except that alumina takes the place of magnesia. Pure kaolin is white, but it is usually colored by impurities, the chief of which are the iron oxides, as in the second specimen, and carbon. It is the basis of all clays and slates; and, although produced in every instance by the decomposition of other aluminous minerals, especially the feldspars, it must usually be regarded as an original as well as an essential constituent of the rocks in which it chiefly occurs — the clays and slates. It is, however, an important accessory constituent or impurity of the limestones, iron ores, coals, sandstones, and other rocks.

Pinite (**44**) is a much less abundant mineral than kaolin, although having a similar origin, resulting, however, from the partial, instead of the complete, decomposition of feldspathic minerals. It is compact and greenish, resembling serpentine, but with a chemical composition, as just stated, between feldspar and kaolin. *Pyrophyllite* (**49**) exhibits similar characters, but contains no alkalies.

Glauconite (**50**) forms the rock greensand, which is chiefly this mineral in rounded grains. It is also a common accessory in other sedimentary rocks. It is especially characterized by its soft, granular or earthy texture and dark green color. Chemically it is distinguished by the large proportion of potash, which makes it a valuable fertilizer.

Micaceous Section. — This section includes the hydrous silicates of alumina and potash or soda having also a highly foliated or micaceous character. The only important species are the hydromicas.

Hydromica (**45**) is the name of a family of hydrous silicates corresponding in composition and physical characters, in a general way, with the micas. The particular species shown is *Margarodite*, but in geology it is not usually important, even when possible, to distinguish the various species, and they are conveniently classed together under the common name hydromica. This is a prominent rock-constituent, and, although often derived from anhydrous species, it may be properly regarded as both original and essential in the rocks in which it chiefly occurs.

CALCAREOUS GROUP. — This very natural group, which ranks next in abundance to the silicates, includes the carbonates of lime and magnesia, the sulphate and the phosphate of lime ; or all of the important compounds of these alkaline earths not belonging to the silicates.

These species are contrasted with most of the silicates by their greater softness and solubility. The carbonates, especially, dissolve freely, with effervescence, in common acids, and dissolve much more readily than most other rock-forming minerals, except halite, in natural waters. With the exception of the crystallized form of apatite, they occur chiefly as both original and essential constituents of rocks ; although the carbonates and sulphates are quite common alteration products, in the rocks, of the basic silicates.

Carbonates of Lime and Magnesia. — The three carbonates of lime and magnesia all crystallize in rhombohedrons, with rhombohedral cleavage, and agree very closely in physical characters.

Calcite, carbonate of lime, is by far the most abundant mineral not belonging to the group of silica and the silicates. It is the principal, and only essential, constituent of all kinds of limestone. Calcite embraces many varieties, only those of greatest geological interest being represented by the specimens. The most typical or perfect calcite is seen in *Iceland spar* (**61**), so remarkable for its strong double refraction. The next specimen (**66**) shows the cleavage equally well, but is less transparent, representing more closely, except in its coarse and perfect crystallization, calcite as it occurs in the crystalline limestones or marbles (**62**). Chalk (**67**), although composed largely of microscopic shells, is essentially an amorphous form of calcite, and, with the admixture of clay and other impurities, represents calcite as it occurs in the compact and earthy forms of limestone. Calcite plays an important role in the organic world, being the principal constituent of the shells (**64**), and skeletons (**68**), of the lower animals. The great geological significance of these organic forms of calcite will be appreciated when it is understood that limestones are formed chiefly through the accumulation of these calcareous organisms (shells, corals, etc.) on the sea-floor.

Calcite is especially characterized by its rhombohedral cleavage, softness, and ready effervescence with cold, dilute acid. The most characteristic microscopic feature is the repeated

twinning, shown in the figure (63), which closely resembles
the repeated twinning of the plagioclase feldspars.

Dolomite (65) the double carbonate of lime and magnesia, is so similar to calcite in most respects that it is
only necessary to note the points of difference.

Dolomite, although a vastly abundant mineral, is much rarer
than calcite, and never occurs in organic forms. It is a little
harder and heavier than calcite, does not exhibit the repeated
twinning structure, and dissolves less readily with acids. It
forms the rock dolomite, of which the specimen is a coarsely
crystalline, pure example.

Magnesite (69), the carbonate of magnesia, is a comparatively
rare species which differs from dolomite very much as that does
from calcite. It is often intimately associated with serpentine,
and appears then as an alteration product of that mineral.
The massive beds of pure magnesite, however, are original chemical deposits from the waters of the ocean. Magnesite is usually quite compact, resembling chalk, but much harder.

Sulphate of Lime.—The hydrous sulphate of lime—
gypsum — is the only one of the numerous native sulphates of special geological interest. The most of the
species in this class are extremely soluble and found only
in solution.

Gypsum (61–62) is chiefly, like dolomite and magnesite, a chemical precipitate from the waters of the ocean,
and hence original and essential ; but it is formed to
some extent, also, by the alteration of limestone and of
the basic lavas. Gypsum is especially characterized by
its softness and lightness ; and it is readily distinguished
from the carbonates by these characters and by its not
effervescing in acid. The clear, crystalline variety *selenite*
(61), remarkable for its perfect cleavage and transpar-

ency, occurs chiefly in veins; while the great beds of gypsum are built up chiefly of the compact or finely crystalline form (62), which, when pure and translucent, is known as *alabaster*.

Phosphate of Lime.—The native phosphates, like the sulphates, are very numerous; but, as before, there is only one species of special geological interest, the phosphate and chloride of lime. This, however, occurs in such diverse forms as to be really equivalent to a group of species.

Apatite (63-64), like calcite and silica, belongs to both the organic and inorganic kingdoms. As a distinctly crystallized, typical mineral (64), it occurs chiefly as an accessory, in veins and in metamorphic and eruptive rocks. Through the decay of these rocks, it becomes a constituent of soils and of natural waters; and in this form it is appropriated by the plants which form the food of the higher animals. This does not explain why the skeletons of the higher animals are chiefly composed of apatite, instead of calcite and silica, as in the lower forms of life, but it shows how the animals obtain the apatite, which not only forms their bones (63), but is also found in their excrement. Both the bones and the excrement may, under favorable conditions, accumulate so as to form important deposits of guano and other forms of phosphate rock, which are so valuable as fertilizers.

HALOID GROUP.—The haloid minerals are mainly soluble in water, and among all the native chlorides, bromides, iodides, and fluorides, only one, *chloride of sodium*, is found in the solid form in sufficient abundance to merit attention here.

Halite or *Rock-salt* (**65**) exists chiefly in solution in the waters of the ocean, of which it forms nearly three per cent. ; and it is found in the solid form in the rocks only where, through the oscillations of the earth's crust, portions of the sea have been detached from the main body and have gradually evaporated to dryness. The most prominent characteristics of halite are its cubic form and cleavage, its softness, lightness, ready solubility, and saline taste.

FERRUGINOUS GROUP.—Iron is so strongly contrasted, especially in its high specific gravity, dark color, and magnetism, with all the other chemical elements of which rocks are chiefly composed, that the mineral species in which it is the chief or only metallic element constitute a very natural geological group. This group embraces the sulphide, several oxides, and the carbonate of iron ; and it is in their origins, associations, and modes of occurrence, rather than in their more strictly mineralogical features, that these can be described as closely related.

Sulphide of Iron.—The sulphides are, as a class, heavy, metallic minerals, of vast economic importance, embracing a large proportion of metallic ores ; but the common bisulphide of iron—pyrite—is the only species of much importance as a constituent of rocks.

Pyrite or Iron Pyrites (**81–82**), is a metallic, pale yellow, hard, heavy species, crystallizing in the isometric system and occurring as an accessory in nearly all classes of rocks, but especially in the slates and the basic eruptive rocks. The crystals are usually either six-sided (cubes) (**81**) or twelve-sided (pyritohedrons) (**82**).

. Although pyrite is hard enough to strike fire with steel, to which circumstance it owes its name (pyrite or fire stone), it is readily decomposed by oxidation to the insoluble oxide of iron or the soluble sulphate of iron; and it is in this way an important cause of the disintegration and discoloration of rocks. In certain rocks, especially in marls and clays, the same chemical substance often crystallizes in orthorhombic forms as the species *marcasite*, which is generally similar to pyrite but decomposes much more readily.

Oxides of Iron.—The various oxides of iron are, after silica or quartz, the only oxides occurring abundantly in the rocks. They are chiefly original and essential, forming solid beds of iron ore; but are also very common accessory constituents in a great variety of sedimentary and eruptive rocks. They are, as a rule, hard, heavy, metallic, dark-colored, insoluble and, unlike pyrite, but little affected by the action of the elements.

Limonite (**83–84**) is the hydrous sesquioxide of iron, containing about 60 per cent. of the metal. Although never crystalline, it occurs in a variety of forms, the purer kinds being either botryoidal and fibrous (**83**) or stalactitic (**84**). It is a common pseudomorph, especially after the sulphide of iron. Its specially distinguishing features are the yellow streak and the absence of magnetism.

Hematite (**85–87**) is the anhydrous sesquioxide of iron, with 70 per cent. of the metal. It crystallizes in flat rhombohedrons having a brilliant metallic luster — *specular hematite* (**85**). The crystals are often flattened to thin scales as in the *micaceous hematite* (**87**). When not distinctly crystalline, the purer forms are commonly botryoidal and fibrous (**86**), resembling limonite. It agrees with limonite, also, in not being attracted by the magnet, but is distinguished by its red streak.

Menaccanite or *Titanic Iron Ore* (**88**) is like hematite with a variable proportion of the iron replaced by titanium. It is very similar to hematite in form and physical properties, except that it is slightly magnetic and the streak is darker. It also never occurs in earthy forms; but is found very generally with magnetite in basic metamorphic and eruptive rocks.

Magnetite or *Magnetic Iron Ore* (**89-90**) is the combined protoxide and sesquioxide of iron, containing 72.4 per cent. of iron. It is the richest of all the iron ores, and the only one containing enough iron to make it strongly magnetic. It is always crystalline (isometric system), and is farther distinguished by a black streak. As an ore, *i. e.*, as an original and essential constituent, it is usually massive and granular (**89**), each grain being an imperfect crystal ; while as an original and accessory constituent it is commonly in distinct and isolated octahedrons (**90**).

Carbonate of Iron. — Mineralogically, the carbonates of lime, magnesia, and iron are very homogeneous, except that the latter is heavier and darker colored than the former ; but geologically we find that the carbonate ore of iron is formed under essentially the same conditions as the uncrystalline oxide ores and is, to a large extent, associated with the same classes of sedimentary rocks.

Siderite (**91-92**) is found in several distinct forms. The crystallized or sparry siderite (**91**), found chiefly in veins, is the purest ; but it is less abundant than the

compact or concretionary argillaceous and carbonaceous
varieties (92), occurring in stratified deposits.

NATIVE CARBON AND HYDROCARBONS.—The hydrocarbons are of organic origin, and the native carbon or
graphite may usually be classed in the same way, since
it has certainly been derived chiefly from the hydrocarbons. This group includes essentially three things—
native carbon or *graphite* (81–82), the *coals* (83–84),
and the *bitumens* (85).

Graphite is crystalline, foliated, soft, black, metallic
and has a very smooth or greasy feel. It is found in
beds and veins, and also as an accessory in various sedimentary and eruptive rocks. The *coals* and *bitumens* are
also chiefly essential, but still, in the carbonaceous slates,
limestones, etc., they must be regarded as accessory.

TEXTURES OF ROCKS.

Texture (*grain*) is a general name for those smaller
structural features of rocks which can be studied in
hand specimens, and which depend upon the *forms* and
sizes of the *constituent particles* of the rocks and the
ways in which these are *united*.

By " constituent particles" are meant, not the molecules of
matter composing the rocks, but those particles or masses,
usually of sensible size, the coming together or development in
association of which has made the rock, as, for instance, the
pebbles in conglomerate, grains of sand in sandstone, crystals

---

[1] These two sections are to be regarded as one, the specimens on each
shelf continuing across from the third section to the fourth, with one set
of numbers; the odd decades on the left, and the even decades on the
right.

CLASSIFICATION OF TEXTURES.

| | The constituent particles of rocks are either | | | |
| | Macroscopic, in which case they are or | | Microscopic, in which case they are, under the microscope, or | |
	Fragments	Crystals	Visible	Invisible.
Primary Textures.	Fragmental.	Crystalline.	Compact.	Vitreous.
Laminated — Banded.	Banded Fragmental.	Banded Crystalline.	Banded Compact.	Banded Vitreous.
Laminated — Schistose or Shaly.	Shaly Fragmental.	Schistose Crystalline.	Shaly Compact.	
Porphyritic.		Porphyritic Crystalline.	Porphyritic Compact.	Porphyritic Vitreous.
Concretionary.	Concretionary Fragmental.	Concretionary Crystalline.	Concretionary Compact.	Concretionary Vitreous.
Vesicular.		Vesicular Crystalline.	Vesicular Compact.	Vesicular Vitreous.
Amygdaloidal.		Amygdaloidal Crystalline.	Amygdaloidal Compact.	
Tufaceous.			Tufaceous Compact.	
Friable or Earthy.	Friable Fragmental.	Friable Crystalline.	Earthy Compact.	

Secondary Textures.

of quartz, feldspar, and mica in granite, etc. As indicated in
the definition, textures may be classified according to the sizes,
the forms, and the modes of arrangement, of the constituent
particles. But of these three modes of division, that based upon
size is the most fundamental, especially if we draw the line at
the limit of visibility of the particles.

Primary Textures of Rocks.

If the particles are macroscopic, they may be divided
according to form ; and undoubtedly the most important distinction to be noted here is that between crystals
and crystalline particles on the one hand, and irregular
or water-worn fragments on the other. Thus, as shown
in the collection and in the table, we arrive at a definition of two of the most important textures—the *fragmental* and the *crystalline.* There are numerous varieties
of each of these, based upon peculiarities in the forms or
sizes of the particles. Thus among the fragmental textures we have the *arenaceous* texture (**1**), the *conglomerate* or *puddingstone* texture (**2**), the *breccia* texture
(**3**), the *shelly* texture (**4**) where the rock is made up of
broken shells, etc ; and the crystalline rocks may be
finely crystalline (**5-6**), or *coarsely crystalline* (**7-8**).

If the particles are microscopic, we cannot, of course
(without using the microscope), divide them according to form ; but, paradoxical as it may seem, they can
be divided according to size. We can distinguish those
rocks in which the particles are merely minute, becoming visible under the microscope, *i. e.*, those rocks in
which the texture is simply very finely fragmental or
very finely crystalline, from those in which the constitu-

ent particles (accidental impurities and inclosures aside) are, for aught that we can determine, of molecular smallness, not being resolved by the highest powers of the microscope. The rocks of the first class, of which common clay (**13**), clay-slate (**11**), black marble (**14**), and novaculite (**12**), are good examples, are called *compact*; and those of the second class, presenting, as in obsidian, etc. (**15–18**), a perfectly continuous and lustrous surface, are termed *vitreous*.

The four textures which we have now defined, and which, it will be observed, are determined entirely by the *forms* and *sizes* of the constituent particles, are called the primary textures; because every rock *must* possess one of them. We cannot conceive of a rock which is neither fragmental, crystalline, compact, nor vitreous; but it is not rare to find two of these primary textures, as the vitreous and crystalline, or compact and crystalline, combined in the same rock.

Secondary Textures of Rocks.

In addition to at least one of the primary textures, a rock may or may not possess one or more of what are called 'secondary' textures. These are determined by the way in which the particles are united, the mode or pattern of the arrangement, etc. The number of secondary textures is rather indefinite; but all the more important kinds are illustrated in the collection. In some rocks, several secondary textures are combined with the same primary texture. The collection and table, however, are constructed to show only the possible combinations of one secondary with a primary texture.

The most important secondary texture is the *laminated*. This exists where the constituent particles, whatever their forms or sizes, are arranged in more or less regular and parallel bands or layers. We may distinguish two principal kinds of laminated texture, according as the lamination is or is not accompanied by easy splitting or cleavage. *Banded* is a suitable name for the texture, in the absence of easy splitting. But, where easy splitting exists, lithologists have found it convenient to distinguish between the easy splitting of crystalline rocks on the one hand and of the fragmental and compact rocks on the other: the term *schistose* being applied and restricted to the former, and the term *shaly* to the latter. We may also properly distinguish between both these and the easy splitting that is independent of, and usually at variance with the layers varying in composition, color, or texture, *i. e.*, slaty cleavage or slaty texture properly so-called.

The specimens on the second shelf not only illustrate these different varieties of the laminated texture, but also like the table, show their relations to the primary textures. Thus the specimens in the first group are fragmental, as their position under that primary texture indicates; and they teach us that fragmental rocks may be either banded (21–22) or shaly (23–24). Similarly, the specimens in the second group teach us that crystalline rocks may be banded (25–26) or schistose (27–28). While from the third group we learn that compact rocks may be banded (33–34), shaly (31) or slaty (32); and the specimens of obsidian (35–36), shows that vitreous rocks may be banded.

We have the *porphyritic* texture when *separate* and *distinct crystals* of *any* mineral, but most commonly of feldspar, are enclosed in a relatively fine-grained base or matrix, which may be either crystalline, compact or vitreous, but rarely fragmental. In all cases of porphyritic texture, the primary texture is, in part at least, crystalline, and it may be wholly so. Nevertheless, crystals alone do not make the porphyritic texture, but that, as with all secondary textures, depends very largely upon the mode of arrangement of the constituent particles. This texture, as stated, occurs commonly in association with the crystalline (**41–42**), compact (**51–52**), and vitreous (**53**), but rarely with the fragmental.

The *concretionary* texture exists when one or more constitutents of a rock have the form, in whole or in part, not of distinct angular crystals, but of rounded concretions, the concretions taking the place in this texture of the isolated crystals in the porphyritic texture. This texture is of common occurrence in connection with the fragmental (**43**) and compact (**54–55**) textures; and the nodules of mica in granite (**44**) show that it may occur with the crystalline texture ; while spherulitic obsidian (**56**) is a good example of its association with the vitreous texture.

The *vesicular* texture refers chiefly to the rounded or bubble-shaped cavities formed by the steam in many volcanic rocks. These steam holes or vesicles occur in vitreous or glassy lavas (**73–74**), very commonly in finely crystalline or compact lavas (**71–72**), and occasionally in the more coarsely crystalline kinds (**63–64**). A somewhat similar texture, which may be called the *cellular* texture, is often developed in fragmental **(61–62)** and compact rocks by the dissolving out of small fossils and crystals.

The *amygdaloidal* texture is produced when, in the course of time, the vesicles of common lava become filled with various minerals deposited by infiltrating waters. The name is from the Latin *amygdalum*, an almond, in allusion to the ellipsoidal form of the vesicles, or amygdules, as they are called after being filled.   The amygdaloidal texture is thus necessarily preceded by the vesicular, and is limited to the same classes of rocks, almost all kinds of amygdaloidal rocks being compact as regards their primary texture **(75–76)**.

The *tufaceous* texture is found in the tufas, or the rocks which, in their most typical developments, have been formed by the deposition of mineral matter from solution over and among moss and other kinds of vegetation.   It bears some resemblance to the vesicular structure ; but as the specimen **(95)** shows, it is essentially a mineral network, reticulated rather than vesicular.

All the secondary textures illustrated so far are determined by the mode, *i.e.*, the pattern, of the arrangement of the constituent particles ; but there are two important varieties depending on the strength of the union of the particles.   These are the *firm* or *strong*, which is illustrated by nearly all the specimens, and the *friable* or *earthy*, which is shown in the remaining specimens on the bottom shelf.   Friable and earthy are not strictly synonymous terms ; but the former is properly restricted to fragmental **(81–82)** and crystalline **(83–84)**, and the latter to compact **(91–92)** and vitreous **(93-94)** rocks.

SYSTEMATIC LITHOLOGY.

In this section we have to illustrate : first, *the Classification of Rocks ;* and, second, *the Descriptions of Rocks or Descriptive Lithology.*

CLASSIFICATION OF ROCKS.

The classifications of rocks which have been proposed at different times are almost as numerous as the rocks themselves. Some of these are obviously artificial, as when we classify stones according to their uses, etc. But we want something more scientific, a *natural* classification ; that is, one based upon the natural and permanent characteristics of rocks. Rocks have been classified according to chemical composition, mineralogical composition, texture, color, density, hardness, etc. ; but these arrangements, taken singly or all combined, are inadequate.

A *natural* classification may be defined as a concise and systematic statement of the natural relations existing among the objects classified. Now the most important relations existing among rocks are those due to their different origins. We must not forget that lithology is a branch of geology, and that geology is first of all a *dynamical* science. The most important question that can be asked about any rock is, not *What* is it made of? but *How* was it made? What were the general forces or agencies concerned in its formation? Rocks are the material in which the earth's history is written, and what we want to know first concerning any rock is what it can tell us of the condition of that part of the earth at the time it was made and subsequently.

The geological agencies, as has been explained in the Guide to Dynamical Geology (p. 27), may be arranged in two great classes : first, the igneous or subterranean agencies, originating in the central or interior heat, and producing the eruptive or unstratified rocks ; and, second, the aqueous or superficial agencies, originating in the solar heat, and producing the sedimentary or stratified rocks.   Hence, we want to know first of any rock whether it is of igneous or aqueous origin.   Then, if it is a sedimentary rock, whether it has been formed by the action chiefly of mechanical, chemical, or organic forces. And, if it is an eruptive rock, whether it has cooled and solidified below the earth's surface in a fissure, and is a dike or plutonic rock ; or has flowed out on the surface and cooled in contact with the air, and thus become an ordinary lava or volcanic rock.

Here we have the outlines of the classification which is illustrated by the collection in sections 5 and 6. Each important type or species is represented by one or more examples ; and the whole may be regarded as essentially an epitome of the main systematic collection of rocks, which occupies sections 7–23, inclusive.   The details of the classification, except so far as they are self-explanatory, may be most profitably considered in connection with the descriptions of the successive groups of rocks in the following pages.

		Unconsolidated.	Consolidated.
Sedimentary or Stratified Rocks — MECHANICALLY FORMED	Conglomerate Group.	Gravel.	Conglomerate.
	Arenaceous Group.	Sand.	Sandstone.
	Argillaceous Group.	Clay.	Slate.
	Volcanic Group.	Volcanic Dust and Sand.	Volcanic Tuff and Agglomerate.

Coal Group.	Iron-ore Group.	Calcareous Group	Metamorphic Group (Silicates).	
			Acidic.	Basic.
Peat. Lignite. Bit. Coal. Anthracite. Graphite. Asphaltum.	Limonite. Hematite. Magnetite. Siderite.	Limestone. Dolomite. Magnesite. Gypsum. Rock-salt.	Feldspathic (Gneisses). Gneiss. Syenite.	Diorite. Norite.
	Siliceous Group. Tripolite. Chert. Flint. Geyserite Novaculite.		Non-feldspathic (Schists). Mica Schist. Hornblende Schist. Talc Schist.	Amphibolite. Chl. Schist. Greensand. Serpentine.

Sedimentary or Stratified Rocks — CHEMICALLY AND ORGANICALLY FORMED

Eruptive or Unstratified Rocks

PLUTONIC or DIKE

This part of the classification is a blank, for the reason that no important eruptive rocks are known which are chiefly composed of minerals belonging to the classes of Native Elements, Chlorides, Oxides, Sulphates, or Carbonates; i. e., nearly all eruptive rocks, so far as known, are principally composed of minerals belonging to the class of Silicates.

Feldspathic		
Granite.	Diorite.
Syenite.	Diabase.

VOLCANIC

Feldspathic	
Rhyolite.	Andesite.
Trachyte.	Basalt.
Obsidian.	Tachylite.
Petrosilex.	Porphyrite.
Felsite.	Melaphyr.

Descriptions of Rocks.

1.— Sedimentary or Stratified Rocks.

1. MECHANICALLY-FORMED OR FRAGMENTAL ROCKS.—
These consist of materials deposited from *suspension* in
water, and the process of their formation is throughout
chiefly mechanical. The materials deposited are mere
fragments of older rocks, worn from the surface of the
land by the agents of erosion ; and, if the fragments
are large, we call the newly deposited sediment gravel ;
if finer, sand ; and if impalpably fine, clay. These
fragmental rocks cannot be classified chemically, since
the same handful of gravel, for instance, may contain
pebbles of many different kinds of rocks, and thus be
of almost any and variable composition. Such chemical
distinctions as can be established are only partial, and
the classification, like the origin, must be mechanical.
Accordingly, as just shown, we recognize three principal groups based upon the sizes of the fragments ; viz. :—

    (1)   Conglomerate group.
    (2)   Arenaceous group.
    (3)   Argillaceous group.

This mode of division is possible and natural, simply because,
as has been explained in the Guide to Dynamical Geology (p.
71), materials arranged by the mechanical action of water are
always assorted according to size. When first deposited, the
gravel, sand, and clay are, of course, perfectly loose and unconsolidated; but in the course of time they may, under the influence of pressure, heat, or chemical action, attain almost any
degrée of consolidation, becoming *conglomerate, sandstone,* and

slate, respectively. The pressure may be vertical, where it is due to the weight of newer deposits; or horizontal, where it results from the cooling and shrinking of the earth's interior or tidal friction, *i. e.*, from the same great causes involved in the formation of mountains, rock-folds, and slaty cleavage. (See Dynamical Geology, pp. 28-30.) The heat may result from mechanical movements, or contact with eruptive rocks; or it may be due simply to the burial of the sediments by newer deposits, which, it will be seen, must virtually bring them nearer to the great source of heat in the earth's interior, on the same principle that the temperature of a man's coat, on a cold day, is raised by putting on an overcoat. The effect of the heat, ordinarily, is simply drying, coöperating with the pressure to expel the water from the sediments; but, if the temperature is high, it may bake or vitrify them, just as in brick-making. Sediments are consolidated by chemical action when mineral substances, especially calcium carbonate, the iron oxides, and silica, are deposited between the particles by infiltrating waters, cementing the particles together. This principle may be demonstrated experimentally by taking some loose sand or gravel and wetting it repeatedly with a saturated solution of some soluble mineral, like salt or alum, allowing the water to evaporate each time before making a fresh application. The interstices between the grains and pebbles are gradually filled up, and the material soon becomes a firm rock. But the student should clearly understand that, in geology, gravel, sand, and clay are just as truly *rocks* before their consolidation as after. It is plain, then, that in each group of fragmental rocks we must recognize an unconsolidated division and a consolidated division.

(1) *Conglomerate Group.*— The rocks belonging in this group we know before consolidation as *gravel*, and after consolidation as *conglomerate*.

Gravel.— The pebbles, as we have already seen, are usually, though not always, well rounded or water-worn ; and they may be of any size from coarse grains to bowlders. The specimens (1-4, 6–7) are typical examples of gravel of fine to medium texture, chiefly from the beaches in this vicinity. Since granitic and felsitic rocks predominate among the harder rocks of this region, they are also, as the specimens show, very prominent constituents of our gravel and shingle beaches.

Conglomerate. — Consolidated gravel. Clay can be very thoroughly hardened by the direct action of heat or pressure ; and when, as in the case of our Roxbury pudding-stone, the paste or finer material of the conglomerate is largely argillaceous, we may fairly suppose that the hardening or consolidation of the rock has been affected mainly through the induration of the clay in which the pebbles are imbedded, by pressure, aided more or less by a higher temperature, the process being illustrated by the specimen of concrete (81), on the bottom shelf, in which mortar takes the place of clay. In other cases, however, where we have beaches of clean pebbles and sand, with little or no clay, it is evident that the consolidation of the rock must be attributed chiefly to chemical action. Several of the specimens (5, 8–9) are clear illustrations of recent formation. In fact, there are many places in meadows, and in the neighborhood of springs, where pebbles are being gradually coated and cemented by iron oxide or carbonate of lime ; and the formation of conglomerate from gravel can thus be actually witnessed. The specimen from Cuba (21), is a recent marine conglomerate cemented by carbonate of lime. It was formed in the immediate vicinity of a coral reef, where the water, flowing over the broken coral, is often saturated with carbonate of lime. The large spec-

imen (27), from Panama, N. Y., is a splendid example
of a recently formed quartzose conglomerate with an
abundant ferruginous cement. Many of the older conglomerates are very imperfectly cemented (22, 26) ; and
it is instructive to compare the pebbles afforded by the
disintegration of such specimens (23), with those still
imbedded in the rock, and with the pebbles now forming
on the beach (6). We recognize two principal varieties
of conglomerate based on the forms of the pebbles. If,
as is usual, these are well rounded and water-worn, the
rock is true *pudding-stone* (26–27 and most of the specimens) : but, if they are angular, or show but little wear,
it is called *breccia* (25, 29–31, 47–49). Pudding-stone is
the prevailing type. It is formed on beaches and bars,
and wherever there is much mechanical movement and
consequent wear ; while breccia is formed in those comparatively exceptional places where angular fragments
detached from the ledges and cliffs by the frost, etc.,
fall, or are carried, into deep or still water, beyond the
reach of the surf, and thus slowly accumulate without
becoming rounded.

Conglomerates are also classified, according to the
nature of the predominant pebbles, as quartz (22–24,
26–30), slate (21), jasper (47–49), limestone (25, 31,
46, 52), granite (44), etc. ; and, according to the nature
of the cementing substance, as calcareous (5, 8, 21, 42),
ferruginous (9, 27, 29), siliceous (24, 47–49), argillaceous (61), etc. The ferruginous cement may usually be recognized by the reddish or brownish color ; the
calcareous cement causes the stone to effervesce freely in
strong acid ; while the siliceous cement gives very strong,

light-colored stones; and the argillaceous cement is known by the dark color and slaty character of the matrix. In the cupriferous conglomerate from Lake Superior (**43**), native copper forms a part of the cement.

On account of its coarse and irregular texture the stratification of conglomerate can rarely be observed in small masses; but the rounded, water-worn form of the pebbles is usually sufficient to show that it is an aqueous or water-formed rock — a consolidated gravel-beach or bar. Similarly, it is always in order to look for fossils in sedimentary rocks, since all parts of the sea are inhabited by animals or plants; but they are rarely observed in conglomerate. The explanation of their absence is found chiefly in the fact that the hard parts, the shells and skeletons, of organisms are unable to resist the violent mechanical action of the surf and marine currents, which wear away even the hard fragments of quartz and granite. The organic bodies are pulverized and completely destroyed by the ceaseless grinding of the pebble or shingle beach. Such fossils as occur in conglomerate are usually in a fragmentary condition, and are well illustrated by the coralline fragments in the conglomerate from Maine (**50**) and by the angular and water-worn fragments of bone in the so-called bone-breccia (**41, 82**).

The jasper breccia from the Huronian rocks north of Lake Huron (**47–49**) is one of the most striking and beautiful of all the varied forms of conglomerate. The third specimen of this rock is from the drift of southern Michigan, showing how these rocks of unique character and limited distribution may be used in tracing the movement of the great ice-sheet.

The large specimen from Nevada (**29**) is a good example of a recently formed breccia, angular fragments of quartzite having become cemented by iron oxide. The mass of recent limestone breccia (**31**) was formed in a quarry where the quarry water trickled over loose fragments of limestone.

The specimens of quartzose conglomerate on this shelf (**22–24, 26–28**) represent very extensive formations in the Middle

States. The limestone conglomerate from Maryland has been used as an ornamental stone under the name Potomac marble.

Specimen (52) is an interesting limestone conglomerate from the Black Hills. The remaining specimens on this shelf represent various minor varieties of conglomerate which are sufficiently explained by the labels. The specimens on the next shelf are in part typical examples of the pudding-stone and breccia so extensively developed in the vicinity of Boston, and in part of the distinctly metamorphic conglomerates observed in connection with some of the older, crystalline formations. The pebbles of the metamorphic conglomerates have been usually, as in these specimens, flattened and elongated by enormous pressure, and the matrix has also become highly micaceous and crystalline.

(2) *Arenaceous Group.* — The conglomerate group passes insensibly into the arenaceous group; for, from the coarsest gravel to the finest sand, the gradation is unbroken, and every sandstone is merely a conglomerate on a small scale.

Sand.— Like gravel, sand may be of almost any composition, but as a rule it is siliceous; quartz, on account of its hardness and the absence of cleavage, being better adapted than any other common mineral to form sand.

The first specimens (1-2) are typical beach sands, consisting chiefly of quartz, but with occasional grains of feldspar, mica, etc. Magnetite is a very common constituent (3) and often occurs by itself, forming the black or so-called iron sands (4). Garnet and other hard minerals are also often present. The specimen (5) from the beach on Marblehead Neck is a nearly equal mixture of quartz, magnetite, and garnet. And nearly pure garnet sands (6) are occasionally found. The older sands,

which have become buried under newer deposits, are
often colored by limonite (7) or hematite (8). The admixture of broken shells and coral gives the calcareous
sands or marls (9).

Sands formed wholly of comminuted organic remains (10) are
not true sands, in the geological sense, since they have not been
formed by the mechanical wearing away of older rocks; and by
their consolidation we obtain limestone and not sandstone.

Sandstone.—Consolidated sand. The consolidation
or hardening of sand to form sandstone, like the consolidation of gravel to form conglomerate, although due to
some extent to the hardening by pressure and heat of
clay that is often mixed with the sand, is effected chiefly
by chemical action, *i. e.*, by the deposition from solution
between the grains of sand of iron oxide, carbonate of lime,
silica, etc. ; and sandstone, like conglomerate, may thus
be made artificially by the percolation through sand of
mineral solutions.

The ferruginous sand from Hingham (11) is partially cemented by the abundant iron oxide ; but a better,
although semi-artificial example is seen in the next specimen (12). The cement in this case is iron oxide resulting from the rusting of an iron spike in the planking
of a wreck which was buried in the sand.

We find in nature every gradation between perfectly
loose sand and the hardest and strongest sandstone.
Some of these intermediate forms, the easily crumbling
or friable sandstones, are represented by the next specimens (21-22, 26). The remaining specimens on this
shelf have been selected as illustrations of the visible

stratification of sandstones. The stratification is indicated by planes of easy splitting or cleavage in the shaly
sandstones (27–28, 31) and by layers of varying color
or texture in the banded sandstones (24–25, 29–30, 32–
33). The specimens from the Black Hills (32–33) are
especially fine examples of distinct and regular banding.

There are many varieties of sandstone depending upon differences in composition, texture, cement, etc. As regards the
texture of sandstones, we may pass by insensible steps from the
coarsest kinds or gritstones, which are really fine conglomerates,
the grains of sand being more properly small pebbles, to varieties like the freestones and flagstones, which are sometimes
almost impalpably fine and pass into slates. Many of these
grades are represented by the specimens. The essential structural similarity of sandstones and conglomerates is very clearly
shown by the drawing (23), which represents a sandstone magnified. The cementing materials are commonly either: ferruginous (iron oxides), giving red or brown sandstones; calcareous, forming soft sandstones, which effervesce with acid if the
cement is abundant; or siliceous, making very strong, light-
colored sandstones.

The first specimens on the third shelf (41–42, 46–47)
are typical examples of ferruginous sandstones. When
the cementing iron oxide is deficient, it is very likely
to undergo segregation, so as to give the stone a
variegated or blotched appearance (43–44, 48–49).
The distinctly calcareous sandstones are well represented
(51–52). In the next two specimens (53–54) the
cement is chiefly kaolin or clay. The first represents the
common flagstone, a fine grained, argillaceous variety
which splits easily into very regular layers or slabs.

A more exceptional cementing substance is seen in the next specimen (50) from the vicinity of the Pitch Lake in Trinidad.   It is saturated with asphaltum, which binds the grains of sand together.   The architectural variety known as freestone (45, 50) is merely a fine grained, light-colored, uniform sandstone, not very hard, and breaking with about equal freedom in all directions.

The gypsiferous and phosphatic sandstones (61, 66) may oe regarded as varieties of calcareous sandstone, phosphate of lime and gypsum partially replacing calcite.

Sandstones, unlike conglomeraes, have often been deposited under comparatively quiet conditions, and are consequently not infrequently fossiliferous.   The next three specimens (67–69) illustrate the prevailing modes of preservation of the fossils, which, except in the newest sandstone, are usually in the form of molds and internal casts.   Sandstones are essentially porous rocks; and the water percolating through the strata gradually dissolves out the fossils, so that we find in all the older sandstones, not the fossils themselves, but the cavities (moulds) which they occupied.

Arkose (62) is essentially a recomposed granite, i.e., a sandstone which is largely composed of the debris of granite—quartz, feldspar and often mica or hornblende—and consequently resembles granite more or less in appearance.

Quartzite.—This name is commonly applied to the older or metamorphic sandstones.   Strictly speaking, the typical quartzite is an unusually hard sandstone, i.e., one in which grains of quartz are combined with an abundant siliceous cement (8 –82, 86-87).   The next specimen (91) is an admirable illustration of the intimate way in which quartzite and common sandstone are often interstratified, the conversion of sandstone into

quartzite in such cases being due mainly to the solution
and segregation of microscopic siliceous shells and other
organisms existing in the stone. The drawings (**83**) representing the appearance of thin sections of quartzite
under the microscope, throw considerable additional light
upon the origin of this rock.

First the siliceous cement is deposited from solution upon the
grains of quartz in such a way as to tend to restore the crystalline form of the quartz; and, second, by the continuance of this
secondary enlargement of the grains the interstices become
completely filled up; and the accurately fitting grains are inter-
locked or dove-tailed together to form a nearly continuous and
homogeneous mass of quartz.

The next specimens (**88–89**) are somewhat micaceous and
schistose, splitting in thin layers and forming the variety often
called quartz schist, which forms a connecting link between
quartzite and mica schist. In fact the quartzites may be classified with the truly metamorphic rocks—the gneisses and schists
—almost as properly as with the sandstones. Itabirite (**84**) is
an interesting variety of quartz schist in which specular hematite takes the place of mica. The specimen from Nantasket
(**90**) owes its induration to the influence of streams of lava,
which flowed over the sand while it was still unconsolidated.

Itacolumite or flexible sandstone (**92** and **13** on the
top shelf) is, perhaps, the most interesting form of metamorphic sandstone. The flexibility is shown to the best
advantage in the more slender upper specimen. This
rock derives its name from Mt. Itacolumi, in Brazil;
and it is an interesting fact that the diamonds of that
country are usually so intimately associated with the
flexible sandstone that it has been called the *mother-stone*
of the diamond.

(3) *Argillaceous Group.*—Just as the conglomerate group shades off, gradually into the arenaceous group, so we find it difficult to draw any sharp line of division between the arenaceous group and the argillaceous, but we pass from the largest pebble to the most minute clay-particle by an insensible gradation. Although clay, like sand and gravel, may be of almost any composition, yet it usually consists chiefly, sometimes entirely, of the mineral kaolin. The fragmental rocks are thus composed principally of two minerals, quartz and kaolin,—the former predominating in the conglomerate and arenaceous groups, and the latter in the argillaceous group. In like manner, while the conglomerate and arenaceous rocks are always visibly fragmental, the argillaceous group is characterized by the compact texture.

Clay.—Pure clay, or kaolin, is white and impalpable, like China clay (**1**) ; but pure clays are the exception. They often become coarse and gritty by the admixture of sand, forming loam ; and they also usually contain more or less carbonaceous matter, which makes black clays or mud (**7**) ; or ferrous oxide, which makes blue clays (**8**) ; or ferric oxide, which makes red, brown, and yellow clays (**2-4, 9**).

By mixing these coloring materials in various proportions, almost any tint may be explained. Clays are sometimes calcareous, from the presence of shells and shell-fragments or of pulverized limestone (**5-6, 10**). These usually effervesce with acid, and are commonly known as marl. It is the calcareous material in a pulverulent and easily soluble condition that makes the marls valuable as soils.

. The residual clays of the South, which are commonly red, are represented by the specimen (**11**) from Trinidad; while the

·next specimens (12-13) are typical examples of the glacial
clays of the North. The fire clays (14) are especially characterized by their freedom from iron oxide and alkaline substances,
or those materials that tend to make clays fusible. They often
underlie beds of coal, and may be regarded as the soil in which
the coal-plants grew.

Slate.—Consolidated clay. We find all degrees of
induration in clay. It sometimes, as every one knows,
becomes very hard by simple drying; but this is not slate,
and probably no amount of mere drying will change clay
into slate; for, when moistened with water, the dried
clay is easily brought back to the plastic state. To
make a good slate, the induration must be the result of
pressure, aided probably to some extent by heat. True
slate, then, is a permanently indurated clay, which will
not readily become plastic when wet. Several of the
specimens on the second shelf are instructive examples
of semi-lithified clays. Two of these (32-33) illustrate
the usual mode of occurrence of fossils in clays, the enclosed shells being undissolved and almost unbroken.
A third specimen (27) is a well-dried residual clay from
the floor of a great limestone cavern. The celebrated
pipe-stone clay of Minnesota (26), is strictly intermediate
between clay and slate, although belonging to one of the
oldest geological formations. The iron oxide in this clay
may act as a cementing substance; but the next two
specimens (22-23) owe their induration to the heat resulting from the burning of beds of lignite. What may
be regarded as the two most normal examples of semi-
lithified clay (21, 31) have probably been consolidated
chiefly by pressure; the first shows the fissile structure so

characteristic of shales, and the second the joint-structure of true slates.

The remaining specimens on this shelf are very typical
examples of slate. They not only exhibit the characteristic textures and colors, but are also very distinctly
stratified. The visible stratification may take the appearance of (1) a distinct banding, due to the alternation of
layers differing in color or texture (**28–29**); or (2) a fissile structure, as in shale (**24, 25, 30**), the rock splitting
readily in thin layers parallel with the bedding. Many
slates, however, are so homogeneous that the stratification is scarcely visible in small specimens, as the remainder of the collection shows.

Few rocks are richer in fossils than clay and slate; and
these prove that they are stratified rocks. The fossils
are not only numerous, but also, as already observed,
exceptionally well preserved (**32–33**).

The argillaceous sediments are deposited in deep and quiet waters, and under conditions so tranquil that even the most fragile
organic remains usually become buried without being broken;
and after their burial are rarely removed by solution, as in the
arenaceous sediments, because the impervious nature of clayey
sediments prevents the free circulation of water. The consequence is that the fossils are not only found perfect in form,
but being, as it were, hermetically sealed, we also find the fossils themselves and not simply their moulds or casts.

The carbonaceous slates and shales are well illustrated
by the next specimens (**41–42, 46–47**) and these also
show, in the carbonized vegetable forms (ferns, etc.),
the source of the carbon. The slate with annelid trails
from Frenchman's Bay (**42**) shows how perfectly even

such slight indications of life are preserved in these fine grained rocks.

Pyrite (**48**) is one of the most common accessory minerals in slate. Through the weathering of the pyritiferous slates and the oxidation of the pyrite, alum—the double sulphate of alumina and iron—is formed, giving the so-called alum slates (**49–50**). The other specimens on this shelf require no special explanation. The older and more altered forms of slate, like the quartzites, afford a more or less gradual passage into the true metamorphic rocks. The most important structural feature of the older slates is the slaty cleavage, which has its best development in the roofing slates (**61–63**, etc.).

These rarely show the stratification distinctly, for it is a remarkable fact that the thin layers into which they split are entirely independent of the bedding. The cleavage has been developed in the slate, subsequently to its deposition, by pressure; and its relations to the stratification are very clearly exhibited in the roofing slates known as ribbon slates (**62**), which show bands of a different color or texture across the flat surface. These bands are the true bedding, and indicate the absolute want of conformity between this structure and the cleavage. The roofing slates are also a good illustration of the variety of colors possible in this class of rocks.

Porcelainite.—This term and *semi-porcelainite* are applied to clay and slate which have been baked or vitrified by heat, so as to have the hardness and texture of porcelain. The specimens from Trinidad and Bohemia are very typical examples of this natural porcelain (**81–82**) and also of the semi-porcelainite (**85–86**). The heat required for the formation of these specimens as well as of the fine series of melted clays from the Bad Lands of the Yellowstone (**89-92**) was afforded by the combustion of beds of lignite interstratified with the beds of clay. The remaining specimens are the product of volcanic heat, deposits of clay or slate having been covered or penetrated by masses of lava or trap.

(*4*) *Volcanic group.*—It is a well-known fact that many volcanic eruptions are more or less explosive in character; and that the lava ejected during these explosive eruptions is largely in the form of dust and fragments. The distribution of these fragmental forms of lava is determined largely by the violence of the eruptions and by the force and direction of the air currents. The larger and heavier fragments usually fall upon the slopes of the volcano, or over the immediately adjacent country, and often to a considerable depth. Thus the Roman city of Pompeii was completely buried, in the year 79, by fragments of pumice from Vesuvius. The finer and lighter material, the volcanic dust, on the other hand, is spread far and wide over the surrounding regions. We have an extreme example in the great eruption of Krakatoa, in 1883, since the dust thrown out by this stupendous explosion was apparently diffused through the entire atmosphere of the earth, and formed a deposit of sensible thickness over a very large part of the earth's surface. Many volcanoes are, like Mt. Vesuvius, composed largely, in some cases almost wholly, of lava which has been ejected in the solid form as dust and fragments, the steepness of the cones increasing with the proportion of the solid ejectamenta.

The fragments vary in size from the finest and most impalpable dust, which floats in the atmosphere for many days or months, to masses weighing several tons. Whether large or small, they are especially characterized by their highly angular forms; by being composed wholly of volcanic materials; and when they have fallen on the land, or beyond the influence of water, fragments of all sizes are mixed together indiscriminately.

The condensation of steam and copious precipitation of water which are important features of many volcanic eruptions, as well as subsequent rains, often wash the lighter volcanic dust or ashes down to lower levels and distribute them in more or less distinct and horizontal layers. In other cases, where the eruptions are submarine, or the fragments fall directly into the sea, they are subjected to the sorting action of water and become mixed in varying proportions with the ordinary sediments accumulating in successive strata on the ocean floor. In still other cases the volcanic materials are washed off the land by rivers and the surf, and, passing thus through the mill of the ocean-beach, the larger fragments, especially, become more or less rounded or water-worn before they are assorted and deposited by the water.

In their origin, the fragmental ejectments are, of course, volcanic; and there are important reasons why they should be classified with the ordinary volcanic rocks, the lavas ejected in a liquid form. Practically, however, it is often more convenient to associate the fragmental volcanic rocks with the mechanically-formed or fragmental sedimentary rocks, since, as explained in the preceding paragraph, it is often difficult or impossible to distinguish between the fragments which have been ejected by volcanic agency from the interior of the earth and more or less worn, assorted, and stratified by water; and those fragments which are wholly of aqueous origin, having been worn directly by the action of water from the ledges of the land. This distinction is not even theoretically possible, on account of the complete blending by admixture on the ocean-floor of the fragments of unlike origin.

Although, as explained, the volcanic fragmental rocks are usually less perfectly assorted than the ordinary mechanical sediments, they may be classified in the same way : first, according to the sizes of the fragments; and, second, according to the degrees of consolidation. It is customary, however, as the labels indicate, to call the unconsolidated materials simply volcanic dust and sand; and the consolidated volcanic tuff and agglomerate.

Volcanic Dust and Sand.— These unconsolidated materials are very commonly known, also, as volcanic ashes. Although this term is misleading, so far as the origin of these rocks is concerned, a glance at the upper shelf in this section will show that it is a good descriptive name. The entire range of texture, from the finest to the coarsest, may be exhibited in the products of a single volcano, as shown by the specimens from Mt. Vesuvius (**1–4**). The coarser of these specimens, as well as those from Germany, Utah, and California (**5–6, 11–12**), illustrate the essentially angular forms of the fragments, the wearing action of water being conspicuous by its absence. As indicated on the labels, it is usually possible, where the characters have not been obliterated by decomposition, to determine the kind of solid lava from which the detritus has been derived. The specimens of pumice dust from old lake basins in Nebraska and Montana (**8–10**) are good examples of fine material which has been transported by air currents and finally arranged in stratified deposits by the action of water.

Volcanic Tuff and Agglomerate.— The specimens on the lower shelves of this section embrace a very typical series of the consolidated volcanic ashes. These are for the most part soft, earthy-looking rocks, which are distinguished from the ordinary mechanical sediments, (sandstone, etc.), among other ways, by the readiness with which they are decomposed on exposure to the weather. It is to this influence, also, that we owe the rather characteristic buff and pink tints of the specimens. The cementing material is usually argillaceous, although, as the specimen from the Sandwich Islands (**31**) shows, it may consist of iron oxide, carbonate of lime, or other substance deposited by infiltrating waters. The fossiliferous specimens (**32, 51**) show very clearly that the materials have been strewn and deposited in some quiet body of water.

The brown specimens on the bottom shelf, from the Sandwich Islands (**81–82**) represent the so-called palagonite tuffs, which are really solid lava which has been speedily and completely decomposed by the acid vapors accompanying the eruption. This shelf also shows some rather ancient and more indurated and metamorphosed tuffs and agglomerates from the vicinity of Boston and elsewhere. The old volcanic rocks of the Boston Basin embrace a very fine series of the fragmental forms, but that the eruptions were largely submarine is indicated by the beautiful stratification often observed (**83–84**) and by the perfect blending of the volcanic with the ordinary water-worn sediments (**85–88**.)

2. CHEMICALLY AND ORGANICALLY FORMED ROCKS. — It has been already explained (p. 81) that from a geological point of view the differences between chemical and organic deposition are not great, the process

being essentially chemical in each case; and since the limestones and some other important rocks are deposited in both ways, it is evidently not only unnatural, but frequently impossible, to separate the chemically from the organically formed rocks. Unlike the fragmental rocks, the rocks of this division not only admit, but require, a chemical classification. Therefore our arrangement will be essentially the same as for the groups of rock-forming minerals, but in the reverse order, thus:—

(1) Coal group.

(2) Iron ore group.

(3) Siliceous group.

(4) Calcareous group.

(5) Metamorphic group (Silicates).

The silicates are placed last, notwithstanding their great importance, on account of their metamorphic character and because they afford the student a more natural passage from the sedimentary to the eruptive rocks, the latter consisting almost wholly of silicate minerals. Most of the silicate rocks are mixed, *i. e.*, are each composed of several minerals; but some silicate rocks and all the rocks of the other divisions are simple, each species consisting of a single essential constituent.

(1) *Coal Group.*— The rocks of this group are entirely of organic origin and include two allied series, which may always be regarded as simply the more or less completely transformed tissues of plants and animals, viz. :— *coals* and *bitumens.*

The chemical and physical conditions favorable for the formation and accumulation of these two series of hydrocarbons have been explained on pages 74–75. It

was shown that, in the coal series, the vegetation is, during the lapse of time, changed in succession to peat,
lignite, bituminous coal, anthracite, and graphite. The
coals, indeed, make a very beautiful and perfect series,
whether we consider the composition, there being a
gradual, progressive change from the composition of ordinary woody fiber in the newest peat to the pure carbon
in graphite; or the degree of consolidation and miner-
alization, since there is a gradual passage from the light,
porous peat, showing distinctly the vegetable forms, to
the heavy crystalline graphite, bearing no trace of its
vegetable origin. The coals also make a chronological
series, graphite and anthracite occurring only iu the
older formations, and lignite and peat in the newer,
while bituminous coal is found in formations of intermediate age.

Peat. — Although not a typical coal, peat is properly
the first member of the coal series. This incipient coal,
or coal in the process of formation, is. well represented
by the specimens on the upper shelf (**1–5**). They show
very clearly that peat is partially carbonized and more
or less comminuted and compressed vegetation, and, to
some extent, that the vegetable forms are chiefly such as
are characteristic of bogs and marshes.

Brown Coal and Lignite — The specimens on the
next shelf (**21–29**) show that these two forms of coal
are essentially similar and that they are intermediate in
character between peat and the most typical coals, such
as bituminous coal, since, while the woody structure is
well preserved externally, as in peat, internally they are
black, lustrous and highly mineralized, as in bituminous

coal. The specimens (**23, 25**) from high latitudes are interesting as indications of a milder climate in those regions at the time when the beds of fossil fuel were forming.

Bituminous Coal. — This is the middle term of the coal series, and as the specimens on the third shelf (**41–46**) show, it is usually very distinctly stratified. The surfaces transverse to the stratification planes are clean and lustrous, showing no vegetable forms or structure; while the surfaces parallel with the bedding are usually dull, soft and often marked by vegetable forms as well as by the woody structure or grain. From a good specimen of bituminous coal we may thus learn two important facts: first, that it is a stratified rock; and, second, that it is of vegetable origin.

The coals in general are extremely infusible as well as insoluble bodies; but the so-called caking coals (**42**) are bituminous coal which softens in the fire. Besides the mineral matter or ash originally in the wood from which the coal was derived, all kinds of coal contain more or less of mechanical sediment such as clay and fine sand which was washed into the peat bog while the coal was slowly accumulating. Cannel coal (**43, 46**), which has a dull and uniform surface, conchoidal fracture, and rarely shows stratification lines or vegetable forms, is also distinguished by a very high percentage of ash. It may be usually regarded as a consolidated carbonaceous mud.

Pyroschist. — From cannel coal it is but a step to pyroschist (**47–49**), which may be regarded as intermediate between coal and slate. It is a coal containing so large a proportion of ash, chiefly in the form of clay, that it is of no value as fuel; or a slate which is so highly carbonaceous that it will almost or quite burn. Hence the name, which means *fire slate*

By extending the series of specimens and giving the composition of each, it could be shown that between the pure coals and the most typical slates containing little or no carbon, there is a perfect gradation. Although the carbonaceous matter of the pyroschists may agree in chemical character with any form of coal; the majority are undoubtedly to be regarded as earthy forms of bituminous coal. The wollongongite (48) from Australia belongs here. While the pyroschists have no value as fuels, they are sometimes the source of valuable hydrocarbons, and are also pulverized to form a black paint, a cheap form of lampblack.

Anthracite. — This is the hardest and most lustrous, *i. e.*, the most highly mineralized, of the true coals, and rarely shows the stratification plainly, or distinct traces of its vegetable origin (61–65). The iridescent tarnish (64–65), giving the variety peacock coal, is due to thin films of iron oxide. Anthracite is highly carbonized vegetation or nearly pure carbon, being represented among artificial products by charcoal and coke. The native coke or carbonite (63) differs from common anthracite chiefly in being bituminous coal that has been changed more rapidly, by volcanic heat derived from a trap dike, and under less pressure.

Graphite. — Except for impurities, such as clay (87), iron oxide, etc., this last member of the coal series is essentially pure carbon ; and it is usually distinctly crystalline (86), the metamorphic form of coal, occurring chiefly in association with such metamorphic rocks as gneiss and crystalline limestone. We need only refer to its uses : in the manufacture of crucibles, pencils, lubricants, etc.

Asphaltum. — The bitumens play a less important part in the lithological classification than the coals, since they are largely liquid bodies, like petroleum. Asphaltum or mineral pitch **(81-85)** is the only important solid bitumen. One of the specimens from the Pitch Lake on the island of Trinidad **(83)** is interesting as seeming to throw light on the origin of the pitch.

(2) *Iron Ore Group.* — The origin of the iron ores through the agency of decaying organic matter has been quite fully explained on pages 76-78, and it was shown there that the physical conditions are essentially the same as for the coals, the concentration of the iron oxide so widely and thinly distributed in nature taking place chiefly in swampy tracts, where the conditions are also favorable for the formation of peat. The principal iron ores also resemble the coals in forming a natural series as regards the composition, consolidation and crystallization, and age. This is especially true of the oxide ores, which are arranged in a column parallel with the coals and will be described first.

In a complete account of the iron ores it would be necessary to recognize and describe other modes of origin besides that indicated above, which may be regarded as simply the most important or usual mode.

Limonite.—The newest forms of limonite, bog-iron ore, and yellow ocher, are well illustrated by the specimens on the upper shelf **(11-15)**. These may be regarded as the incipient iron ore; and it will be seen that they resemble peat in their porous and earthy texture as well as in their mode of occurrence. The first two specimens **(11, 14)** are very typical examples of the bog ore, but the specimens from the Katahdin Iron Mine (**13, 15**) are especially interesting on account of preserving distinct impressions of the decaying vegetation of the marsh. Where the vegetation is less abundant and more clay is washed into the marsh and mixed with the iron oxide, the yellow ochre (**12**) is formed. The specimens on the second shelf include, besides additional examples of fos- siliferous limonite (**31–32**), some of the more solid and compact forms (**37-38**), as well as one specimen (**33**) to represent the botyroidal, stalactitic and other forms so interesting from the mineralogical point of view. Although these are all regarded as simply old and more or less consolidated bog-iron ore ; yet it is known that these forms are also developed by the oxidation, in place, of pyrite, the sulphide of iron. Limonite is an important ore of iron, containing, when pure, about 60 per cent. of iron and 15 per cent. of water.

Hematite.—This rock, the most abundant of all the iron ores, is believed to result in most cases from the expulsion of the water from limonite by heat and pres- sure. When pure it is a very rich ore, yielding 70 per cent. of iron. The fossiliferous specimens on the second shelf (**34–36**), representing the celebrated Devonion bed at Nictaux, Nova Scotia, and the persistent and valuable oölitic bed in the Clinton formation of the United States, suffice to show that these ores are stratified rocks. They are not, however, wholly original deposits, for it is certain that the fossil shells were once calcareous, and that they have been gradually replaced by iron oxide due, probably to the oxidation of pyrite in overlying strata. The concretionary or oölitic (**54**) and the argillaceous or slaty (**55**) ores on the next shelf are other important varieties of the uncrystalline or earthy hematites; while the remaining specimens illustrate the different grades of the crystalline or specular ores. Among these the itabirite (**52-54**) is of especial geological interest. It is essentially a schist composed of quartz and crystalline scales of hematite instead of mica. Closely related to this is the banded jaspery hematite or jaspilite (**79**) from Marquette, in which the red layers are jasper and the black specular hematite. A large and exceptionally fine specimen of this beautiful rock adorns the driveway on the Newbury street side of the Museum. The magnetic hematite (**59**) affords a natural passage to the next species.

Magnetite and Menaccanite.—These are the oldest and most highly crystalline or metamorphic of the iron ores, important deposits of these ores occurring only in the oldest geological formations. They are (**71-77**) found in great abundance; and magnetite is a pure and valuable ore, the richest of all the iron ores, containing 72 per cent. of metallic iron when pure; while menaccanite (**73**), the double oxide of iron and titanium, rarely contains more than 40 per cent. of iron, and this small proportion is rendered quite valueless, as a rule, by the titanium. While there can be but little doubt that these are usually stratified rocks, their origin is not by any means as clearly established as that of hematite and limonite.

Manganese Ores and Emery.—The manganese ores form almost as distinct a series, as regards composition, crystallization, and age, as the iron ores; and the

newer forms, at least, such as wad (**94**) (bog manganese and manganese ocher) and psilomelane (**95**), appear
to be formed under essentially the same conditions as the
newer iron ores. They are, however, of much rarer
occurrence and of little lithologic importance. Emery
(**96**) appears to occur in bedded deposits and may be
conveniently classed with magnetite.

Siderite.—This ore is carbonate of iron, containing
about 43 per cent. of iron, when pure. In physical
characters and conditions of occurrence, as well as in
composition, it stands somewhat apart from the series of
oxide ores, holding a relation to these analogous to that
of asphaltum to the coal series. In its sedimentary
forms, siderite is usually compact and highly argillaceous (**91**) or carbonaceous (**92-93**). The argillaceous ore or clay-iron-stone occurs very generally in the
form of concretions, often of large size; while the carbonaceous variety or black-band ore is found in more
continuous beds.

We have already learned (p. 78) the reason for the intimate
association of siderite with beds of coal, and this accounts
equally for the carbon contained in the ore itself. The association adds much to the value of siderite as an ore of iron,
since it usually insures its proximity to both the fuel and the
flux (limestone) required for the reduction of the *iron*.

(3) *Siliceous Group.*— We properly include here,
neither the rocks composed chiefly of silicates, like the
gneisses and schists, nor the more truly siliceous rocks
of mechanical origin, like sandstone and quartzite; but
only those rock forms of quartz and opal which may be

regarded as either (1) chemical precipitates, or (2) as due to the accumulation of siliceous organisms. We have thus indicated that this group embraces two sub-groups, according as the origin of the siliceous deposits is distinctly organic or distinctly chemical. The first sub-group includes tripolite or diatomaceous earth, flint, and chert, and the second, geyserite or siliceous tufa, novaculite, and buhrstone.

Tripolite or Diatomaceous Earth.—This interesting rock is well illustrated by the specimens on the upper shelf (**1-8**). It is soft, light, and looks like clay ; but it is lighter. Notwithstanding its softness, it is really composed of a hard mineral, viz., silica in the form of opal. By rubbing off a little of the dust and examining it under the microscope, we could easily prove that the silica is of organic origin, for it would appear to be a mass of more or less fragmentary organic remains, occurring in great variety and of wonderful beauty and minuteness (**3**). There are few rocks so unpromising on the exterior and yet so beautiful within.

We have seen in the Guide to Dynamical Geology that these siliceous organisms are principally the cases of Diatoms, the shells of Radiolaria and the spicules of Sponges; and we can form some idea of their minuteness from Ehrenberg's estimate that a single cubic inch of pure tripolite contains no less than 41,000,000,000 organisms. It has also been shown that these organisms are very widely distributed, in fresh water as well as in the ocean. Tripolite is white (**1-2**) when pure, and light enough to float in water; but by the common admixture of clay and carbonaceous matter it becomes much heavier and darker (**4-6**). The California specimen (**1**) is beautifully laminated. The specimen from Waltham (**6**) is from one of the many bogs in which tripolite is now forming; while the specimens from Maryland and Virginia (**7-8**) represent an extensive stratum in the Tertiary formation. Tripolite owes its value as a polishing material to the fact that it consists of a hard mineral in an exceedingly fine state of division. While its non-conducting and fire-proof qualities give it other important uses in the arts..

Chert and Flint.—Tripolite is not found in the older formations; but during the course of geological time the organic and opaline silica is gradually consolidated and changed to quartz, chiefly by percolating waters, which are constantly dissolving and re-depositing it; and, finally, in the place of a soft, earthy rock we get a hard, flinty one which we call *chert*, if it occurs in the older or *flint*, if it occurs in the newer formations. Besides forming beds of nearly pure silica, which we call tripolite, the microscopic siliceous organisms are diffused more or less abundantly through other rocks, especially chalk and limestone. In such cases the consolidation of the silica implies its segregation also; *i.e.*, the silica dissolved by percolating water is deposited only about certain points in the rock, building up rounded concretions or nodules. Thus, a siliceous limestone becomes, by segregation of the silica, a pure limestone containing nodules of chert, which are usually arranged in lines parallel with the stratification. The specimens illustrate the nodular or concretionary forms of both chert (**34**) and flint (**37-38**).

The distinctly stratified or laminated structure (**21**) and the cuboidal fracture (**22**) are somewhat exceptional. Typical buhrstone (**28-29**) is a distinctly porous or cellular chert, the cellular texture being usually due to the subsequent solution of calcareous shells which were deposited with the microscopic siliceous organisms.   The next specimen (30) is very highly fossilliferous, and shows the possibility of a passage from chert into ordinary shell limestone.   The flint owes its darker color to the greater translucency and to the less perfect oxidation of the carbon originally present in all of these rocks.

Menilite (26–27) is similar to chert and flint, except that it is found chiefly in a still newer formation—the Tertiary—and is, mineralogically, to be classed with opal instead of quartz. *Semi-opal* (33) is much the same, except that it is distinctly stratified instead of nodular.

Geyserite or Siliceous Tufa.—As the name implies, this rock is formed chiefly about hot springs and geysers.   The highly heated subterranean water, as explained on page 48, is usually more or less alkaline through the decomposition of various alkaline silicates at great depths in the earth, and is thus a natural solvent for the silica liberated by the same process.   As the thermal water approaches the surface, its temperature and pressure, and hence its solvent power, are diminished and a large part of its dissolved silica is deposited around the outlet as a snow-white porous tufa. The specimens (41–44, 46–49, 51) are fine examples from the geysers of the Yellowstone National Park ; and exhibit well the various interesting forms naturally assumed by deposits of this character.   Silica deposited in this way, like organic silica, is chemically the same as opal ; and, like tripolite, this rock is always of recent origin, tending with the lapse of time to become more solid and quartzose.

Novaculite.—This is the old siliceous tufa which has

been consolidated and changed to quartz by the action of pressure and percolating waters. The first specimen (45) represents the remarkably fine deposit in Arkansas, commonly known as Arkansas stone and largely used for oil stones, streakstones, &c. It is a white and perfectly compact form of pure silica.[1] The green specimen (50) from Pelham, Mass., appears to be essentially similar, but it is somewhat brecciated. It was formerly used for gun-flints.

(4) *Calcareous Group.*—This group properly includes not only the rocks composed of the calcareous minerals —the carbonates—of lime and magnesia, and the sulphate of phosphate of lime—but also rock-salt, which belongs, mineralogically, to the haloid group. This classification is necessary on account of the intimate dynamic and structural relations of these various rocks, due to the fact, as explained on pages 72-73, that when of purely chemical origin, they are formed in a definite and unvarying sequence during the complete evaporation of a detached portion of the sea. Having regard either to this natural sequence or to the relative abundance of these rocks, the series properly begins with limestone and ends with phosphate rock;. but it has been found necessary to reverse this arrangement in order to better adapt the collection to the cases and to bring the specimens requiring the best light—the marbles—next to the window. Although the limestones are largely of organic origin, this

---

[1] Recent investigations indicate that the Arkansas stone is probably of organic origin, and hence to be classed more properly with chert than novaculite.

group as a whole embraces, so far as we know, the largest and most important series of rocks having a purely chemical origin.   But they are not alone of scientific interest; for from the standpoint of economic geology they easily rank next to the coals and iron ores.

Phosphate Rock.—This species embraces all the stratified rocks which are largely composed of phosphate of lime—the mineral apatite.   They are all, so far as the phosphatic material is concerned, wholly of organic origin, the phosphate being derived either from the excrement or the skeletons of animals.   The series properly begins with the typical guano (**61-62**).   This is simply the accumulated excrement of certain marine birds which subsist upon fish and immense numbers of which inhabit small coral islands in nearly rainless portions of the tropics.   In some cases the coral rock upon which the guano rests is itself changed to phosphate of lime by the infiltration of the more soluble portions of the guano (**63-66**).   The specimens from the Carolinas (**64 65**) represent the important class of phosphate deposits resulting usually from the leaching out of the more soluble portions of phosphatic limestones and the concentration of the phosphate of lime.

Rock-salt.—This interesting and useful rock, as we have already learned, is deposited in a purely chemical way, and only in detached and drying up portions of the sea.   The first specimen (**67**) represents the exceptionally pure deposit in the delta of the Mississippi, and the next specimen (**68**) the bed of great thickness and purity recently discovered at a considerable depth in western New York.   The celebrated salt beds of Cheshire, Eng.,

(69) are less pure, containing some clay and iron oxide ; and from these there is a gradual passage into deposits that are more properly described as saliferous sand and clay, or sandstone and slate, from which the salt is obtained in the form of brine which is evaporated by natural or artificial heat, as when salt is obtained from sea-water.

Gypsum.—When pure this rock is identical with the mineral gypsum (81-85) except that it is rarely distinctly crystallized. It is formed under approximately the same conditions as rock-salt, except when resulting from the alteration of limestone (86) ; and like rock-salt it is usually more or less mixed with other kinds of sediment, especially clay (82-83). It is usually very massive, the distinct stratification of the next specimen (87) being quite exceptional. The very pure and somewhat crystalline forms of rock gypsum are represented by the beautiful specimens from Michigan and Mexico. Gypsum is largely used, under the name of *plaster*, in the manufacture of fertilizers; and these pure forms are also adapted to its higher uses in the manufacture of crayons, plaster of Paris, &c. The solubility of gypsum is well illustrated by the erosion of the large specimen from the Gulf of St. Lawrence (88).

Magnesite.—This is the massive form of the mineral magnesite, and an unimportant rock. It is used in certain chemical manufactures, and in the construction of blast furnaces. The principal source of the commercial supply is the Grecian Archipelago (1).

Dolomite.—This is the rock form of the mineral of the same name and an important constituent of the earth's crust. Like gypsum and rock-salt, dolomite is probably never deposited in the open ocean, but only in closed basins; and in consequence it agrees with them in being rarely fossiliferous. There is a wide range in texture from compact to coarsely crystalline. The specimens on the second shelf (21-25) are fair examples of somewhat impure, compact to finely crystalline dolomite. The specimen from Missouri (23) is interesting on account of the associated lead ore (galenite) and barite; while the specimen from Niagara Falls (25) contains good crystallizations of the mineral dolomite. The semi-crystalline dolomites afford some of the most beautiful and durable marbles. Two of these, from Vermont, the Plymouth and Winooski marbles, are represented by specimens on the next shelf (41-46). They are fine examples of the mottled and brecciated structures in the calcareous rocks. The Winooski marble is especially varied and beautiful, and attention is called to the fine series of polished specimens in section 15, next to the window. The specimens on the lower shelves illustrate the purer and more crystalline forms of dolomite; and it is evident at a glance that many kinds of white marble must belong here (61-68). Some of these specimens, as the labels show, represent the very frequent blending or intermixture of dolomite and limestone.

The specimen from Lee, Mass. (1) is interesting on account of the slender crystals of tremolite, the calcium and magnesium carbonate having been partially changed to the silicate. In like manner we find that the compact dolomite of Stoneham (63) contains some hydrous silicate of magnesia—serpentine. The dolomites of Bolton, Chelmsford, etc., (81-83) are highly crystal-

!ine and quite pure, except for the associated crystallized minerals, but they are excelled in both respects by the so-called
" snow-flake" marble from Pleasantville, N. Y. (84)   This is a
remarkably coarse and perfect dolomite. The dark specimen
(85), which is a weathered piece of impure limestone interstratified with gray bands of dolomite, shows the dolomite layers in
relief on the weathered surface, and thus illustrates both the
more granular texture and the greater hardness and insolubility
of dolomite as compared with limestone.

Limestone.—This is the lithologic form of carbonate
of lime or calcite and one of the most abundant, interesting and useful of all rocks. The composition, including both the essential constituent—calcite (1), and the
principal accessory constituents (2), is well illustrated
by the smaller specimens on the upper shelf. It has
been explained on page 81 that the limestones are
partly of organic and partly of purely chemical origin.
The formation of the first class is well illustrated by the
large piece of coquina (3) on this shelf and also by the
finer coquina (26) and the eolian or wind-blown limestone (27) on the second shelf. The varying textures
represent simply different degrees of comminution of
shells and coral by the action of the waves.   The limestones known to be of purely chemical origin are best
represented by calcareous tufa (28) and travertine (29).
These are formed by the deposition of carbonate of lime
by springs and streams, chiefly in limestone countries,
where the underground waters are often saturated with
carbonate of lime. The organically formed limestones
are usually conspicuously fossiliferous, and the purer kinds
are farther illustrated by the smaller specimens (21-25)

on this shelf. The most of these are of quite recent origin, one specimen (23) showing the coquina still unconsolidated, an organic sand. This part of the illustration is continued by the specimens on the third shelf. These are, however, on the whole, less pure than the preceding, some of the specimens being distinctly argillaceous. There are noticeable and noteworthy differences in the character of the fossils, shells predominating in some specimens (41–42) and corals (44, 50) or the remains of crinoids (52) in others; but the most interesting specimens are, perhaps, the nummulitic limestone (47) from the Tertiary formation of Egypt, the indusial limestone filled with the tubes of caddis worms (45), and the chalk (48) from the Cretaceous formation of England. The beautiful spiral and coin-shaped fossils of the first are readily apparent; while the chalk appears quite unfossiliferous. Microscopic examination, however, shows that it is chiefly composed of the very minute shells of Foraminifera, resembling in structure the tripolite among the siliceous rocks. The soft calcareous ooze which is now forming over extensive areas of the floor of the deep sea may be regarded as a modern chalk deposit. The drawing (43) shows the appearance of the ooze when highly magnified.

The specimens on the fourth shelf may also be regarded as wholly or almost wholly of organic origin; but the comminution and solution of the organic remains during their accumulation has been carried to the point of complete obliteration; and the calcareous detritus has become mixed, at the same time, with a large proportion of argillaceous impurity or clay, so that all of these specimens might be properly described as argillaceous in composition and compact in texture. Among these impure types is included the hydraulic limestone (**63**, **68**), from which the hydraulic cement or water-lime is obtained. The Caen stone (**73**) so extensively used as a building stone in France, the thin-bedded limestone associated with the lithographic stone at Solenhofen, Bavaria, (**69**) and the beautiful specimen from the Mount of Olives (**64**) are fine examples of compact and more or less chalky limestones. But the compact limestones appear to reach their highest perfection in the lithographic stone (**65**, **70**). The best quality comes from Solenhofen, and is remarkably fine and homogeneous. The siliceous or cherty limestones (**71-72**) are very abundant in some of the geological formations. In the first specimen the chert is regularly interstratified with the limestone, forming the dark layer in the rock. The specimens from Ireland on the bottom shelf (**81-82**) are interesting examples of local metamorphism, since these hard, compact rocks have resulted from the alteration of chalk by the heat of eruptive masses which have broken through that formation.

The specimens on the narrow shelf (**91-93**) represent the magnesian limestones, which are usually somewhat crystalline and afford a gradual passage into dolomite. The large specimen (**90**) is another magnesian limestone, the magnesia being in this case largely in the form of the hydrous silicate—serpentine.

The finely concretionary or oölitic and pisolitic limestones of different ages are well represented by the remaining specimens (**84–89**) on this shelf. The first two specimens (**84–85**) illustrate the formation of the concretionary structure. The specimen from Barbadoes (**85**) is simply a hardened calcareous sand, like the eolian limestone from Bermuda (**27**), with somewhat angular grains. But before their consolidation the individual grains, as they are rolled about by the waves, are often coated by successive layers of carbonate of lime, each minute grain thus becoming the nucleus of a little spherical concretion (84) ; and the subsequent cementation of these produces the typical oölite.   The drawing (83) represents a section of oölitic limestone highly magnified, showing the rounded forms and concentric structure of the concretionary grains.   Calcareous tufa has been exhibited as an example of chemically formed limestone (28) ; but this interesting variety is much more fully illustrated by the specimens on the upper shelf of section 15.   The carbonate of lime may be deposited so as to build up more or less solid or earthy masses, as in the smaller specimens (1-4) : or it may incrust moss (7), leaves (6), or roots, branches and other forms of vegetation (5), the organic matter subsequently decaying away.   The thinolite tufa (8) from the Great Basin is especially interesting on account of its pseudomorphic origin, slender crystals of some more soluble salt, due to the evaporation of saline lakes, having been subsequently replaced by carbonate of lime.

The specimens on the second and third shelves represent the limestone marbles, the dolomite marbles, as already explained, occupying the fourth shelf.   The marbles, properly defined, are those varieties of limestone and dolomite capable of receiving a polish an l otherwise adapted to being used as ornamental stones.   The limestone marbles may be quite compact and distinctly argillaceous and carbonaceous (31-32), or mottled and clouded (33-34).   They may be fossiliferous ; and several distinct varieties are based upon the character of the fossils, as encrinal marble (36), shell marble (29-30), etc.   Some of the specimens (26-28) show a gradation through forms in which the fossils are

more finely comminuted, to the compact marbles. ut
the black marble from Isle La Motte (**21–25**) represents
the opposite extreme, the fossils being isolated, large
and perfect.

The first specimens on the third shelf represent the
pure, white, and crystalline—the typical—marbles, both
Italian (**41–42**) and American (**46–47**). These pass into
gray and clouded forms (**51–55**) ; and many varieties,
although distinctly crystalline, owe their beauty and interest chiefly to the colors imparted by iron oxide and
other impurities, just as with the Winooski dolomite
marble on the fourth shelf. This is true of the Lisbon
marble (**43, 57-59**) and the Tennessee marble (**44, 48–
50**). The serpentinic limestone from Port Henry
(**45, 56**) gives an entirely distinct variety of marble,
determined by this interesting accessory mineral.

The large specimen (**84**) on the bottom shelf is a fine
example of the coarsely crystalline or sparry limestones ;
while the other specimens afford additional illustrations
of the more finely crystalline or saccharoidal kinds, the
typical metamorphic limestones, some of them containing graphite, chondrodite, phlogopite, and other characteristic accessories. The drawing (**91**) shows the most
characteristic appearance, in polarized light, of the crystalline limestones and marbles, the parallel stripes and
lines of color being due to repeated twinning and resembling the twinning striae of plagioclase (page 95).

(5) *Metamorphic Group* (*Silicates*). —The rocks of the
preceding groups are nearly all simple, *i.e.*, each consists
of only one essential mineral ; but most of the rocks in
this great group of silicates are composite or mixed, con-

sisting each of several essential minerals. Quartz and magnetite are, however, the only important constituents of these rocks which are not, strictly speaking, silicates; and they are not really exceptions to the rule, since quartz may usually be regarded as an excess of acid in the rock and magnetite as an excess of base. This group of stratified rocks is of exceptional importance, first, on account of the large number of species which it includes, and, second, on account of the vast abundance of some of the species. These are, above all others, the rocks of which the earth's crust is composed. With unimportant exceptions, all the rocks of this group are crystalline; and they constitute the principal part of what is generally included under the term metamorphic rocks—a general name for all stratified rocks which have been so acted upon by heat, pressure, or chemical forces as to make them crystalline. The crystalline limestone, . dolomite, iron ores, etc., show us, however, that metamorphic rocks are not wanting in the other groups.

Some of the details of the classification of this group, as shown in the Table (p. 122), require explanation. In describing the silicate minerals it was stated to be important to recognize two classes—the acidic and the basic, the dividing line falling in the neighborhood of 60 per cent. of silica. Now this division is important simply because Nature has in a great degree kept the acidic and basic minerals separate in the rocks. There is, of course, no sharp line of demarcation between these two great classes of rocks; but the vertical broken line in the table separates those kinds commonly classed as acidic from those commonly classed as basic.

The horizontal line in the table separates the rocks containing feldspar as an essential constituent, from those in which feldspar is wanting except as an accessory constituent. The feldspathic species include gneiss and certain more basic rocks related to it; and they are known in general as the gneisses. While the non-feldspathic kinds, having commonly a schistose texture, are known collectively as the schists. The gneisses properly come first, occupying the first and second shelves in sections 16, 17 and 18, the more acidic species (gneiss and syenite) being placed toward the left, and the more basic species (diorite and norite) toward the right.

Gneiss. — This is the most important of all stratified rocks. It forms not far from one half of New England, and a very large proportion of the earth's crust. The specimens begin with illustrations of the composition, the minerals on the first tablet (**1**) being the two essential constituents of gneiss — quartz and feldspar; while the second tablet (**2**) shows, first, the leading accessory minerals — the black and white micas and hornblende, which form the principal varieties or sub-species of gneiss, and, second, some of the less prominent accessories, occurring in subordinate but interesting varieties.

The remaining specimens on this shelf are good examples of the simplest forms of gneiss, as regards composition — the so-called binary gneiss (**3–7**), consisting almost exclusively of the essential minerals, quartz and feldspar. These binary gneisses are often very obscurely stratified, resembling granite (**5–6**), and contain fewer accessory minerals, such as garnet, etc., than the gneisses which hold an abundance of mica.

The distinctly micaceous gneisses constitute the leading and specially important variety; and these are well represented by the specimens on the next shelf. Some of these (**21–22**, **26–27**) are very highly micaceous, approaching the mica schists in both composition and schistose texture ; while in other cases the mica is either so deficient or so fine that its influence on the texture is inappreciable, and the gneiss appears granitoid (**23, 28**), as in the case of the binary gneiss. It is probable that some of the so-called gneisses really are true granites, the flow-structure or superinduced foliation of an eruptive rock being mistaken for stratification.

Passing along this shelf into the next section (17) we come to some exceptionally typical examples of granitoid micaceous gneiss (**21–22**). The granitoid gneisses, especially, are often valuable building stones. They are, however, usually quarried and used under the name of granite. The remaining specimens on this shelf sufficiently represent the hornblendic gneisses, in which hornblende partly or wholly replaces the mica in the preceding variety. The specimens from Shelburne Falls and Northfield (**23–24**, **30**) are very plainly stratified ; but the example from the Highlands of the Hudson (**29**) is in other respects the most typical. The varieties on the upper shelf of this section (17) are especially interesting on account of the accessory minerals. These include cyanite (**1**) ; iolite or dichroite (**4**), forming the variety dichroite gneiss ; and chlorite (**2**), the granitoid and chloritic gneiss forming the main part of the Mt. Blanc Range, having long been known as *protogine* (**3**). The highly garnetiferous binary gneiss (**5**) is peculiarly interesting on account of the high latitude from which it has come.

Syenite, Diorite and Norite. — The specimens on the first and second shelves of the next section (18) represent these more basic gneissoid rocks. They are, even taken all together, far less abundant and less varied, as well as economically less important, than the true

gneisses ; and for these reasons, and also because they
are regarded by many geologists as belonging wholly to
the class of eruptive rocks, the illustration is comparatively limited.   The composition of the different varieties is expressed with sufficient fullness on the labels.
The two specimens of syenite from Marblehead (**1, 4**)
are quite typical ; but it occurs there under conditions
which make its sedimentary origin doubtful.   Neither of
the diorite specimens are well characterized ; the first (**2**)
being too highly hornblendic and the second (**3**) too
compact and slaty.   The norite specimens on the second
shelf are all from the west shore of Lake Champlain and
are fine illustrations of the principal varieties of this
rock.   Some (**21**) are nearly pure Labrador feldspar,
while others show the supposed stratification (**24**) very
plainly and also an interesting accessory mineral —
garnet (**22**).

  Mica Schist. — This is by far the most important, as
well as the most typical, of all the schists ; and, next to
gneiss, it is the most abundant rock in New England.
The illustration begins, as before, with the composition.
First, the essential constituents — quartz and black and
white mica (**42**) ; and, second, a few only of the numerous and interesting accessory minerals (**44**).   The first
two specimens of mica schist (**41, 43**) may be regarded
as very typical examples, exhibiting the schistose texture
to good advantage.   One of them (**43**) shows in addition
garnet, the most important crystalline accessory of mica
schist.

  The remaining specimens on this shelf (**45–49**) are
examples of highly siliceous or quartzose schists, contain-

ing only enough mica to develop the schistose texture. They indicate a gradual passage which is often observed from mica schist to quartzite. The specimens on the next shelf include the beautiful schist with green mica, from Maine (61); and, besides additional examples of garnetiferous mica schist, varieties depending upon other accessory minerals, such as tourmaline (67), fibrolite (68–69), and ottrelite (70).

The illustration of the varieties of mica schist is continued on the third shelf of the next section (17). We find here a good specimen of staurolitic schist, containing distinct cruciform twins of staurolite (49). The crystals are, however, often much less conspicuous (50). Very similar to the staurolitic schists, and scarcely less abundant, are those containing andalusite (45) and especially the impure variety of andalusite known as chiastolite (46–47).

The chiastolite is readily recognized by the appearance of a black square or cross on the sections of the crystals (see p. 103); and both the staurolitic and chiastolitic mica schists are usually highly argillaceous and of a distinctly slaty structure. Some of the plain argillaceous schists (41), especially, show very clearly how we may have a perfect transition from mica schist into common slate. These mica schists in which kaolin takes the place of quartz are always imperfectly crystalline and slaty, and sometimes calcareous (42) or carbonaceous (48). Pyrite is one of the less important accessory minerals (51). The important variety hydromica schist, in which hydromica takes the place of ordinary mica, is well illustrated by the specimens (62-64, 66–68) on the fourth shelf of this section (17). These are distinguished by being softer, greener, and less distinctly crystalline; and they are also often, like the true mica schists, argilla-

ceous (68), and contain other hydrous silicates, such as chlorite
and pinite (64), and they may also be calcareous (66). Garnet
and other anhydrous accessory minerals occur much less frequently than in true mica schist.

Hornblende Schist. — This rock is like mica schist
with hornblende in the place of mica. It is a less typical schist, and also much less abundant, than mica schist.
The composition is most clearly illustrated by the coarsely
crystalline specimen from Chester (45) (3d shelf, section
18), bladed crystals of hornblende as well as the finely
granular white quartz being clearly exhibited. The
finely crystalline kind (41) is, however, much more
normal and abundant. The specimen from Rowe (46)
is a particularly good illustration of the stratification.
The garnetiferous specimen (42) may be regarded as the
exception that proves the rule that hornblende schist is
less characterized by accessory minerals than mica
schist.

Amphibolite or Hornblende Rock.—This rock is
essentially pure hornblende, or like hornblende schist
without the quartz, or diorite without the feldspar, and
gradations are observed between it and both of these
species. The specimens from Chester (43–44) illustrate
the typical variety as well as the gradations.

Schorl Schist, Garnet Rock, etc.--The rarer kinds
among the anhydrous schists are represented by the
specimens on the next shelf (4th shelf, section 18).
These are : (1) *Schorl schist* or *tourmaline schist*, consisting of black tourmaline imbedded in granular quartz.
The specimen from Warwick (61) is well characterized.

(2) *Topaz schist*, consisting essentially of topaz and quartz (**65**). (3) *Garnet rock* (**62, 66**), in which common garnet is the chief and sometimes the sole constituent. (4) *Epidote rock* (**63, 67**), consisting chiefly of massive epidote with more or less quartz, etc. (5) *Graphite schist* (**64**), composed of quartz and graphite, resembling a mica schist in which scaly graphite takes the place of mica. This rock is of economic interest, being an important source of commercial graphite. (6) *Eklogite*, a rock of exceptional and variable composition and of doubtful sedimentary origin (**68, 70**).

The remaining schists, occupying the bottom shelf of sections 16, 17 and 18, include as essential constituents only minerals belonging to the hydrous silicates, such as talc, pyrophyllite, pinite, glauconite, serpentine, and chlorite and we might properly add hydromica to the list.

Talc Schist or Steatite.—This interesting rock is essentially pure talc, though often containing quartz, chlorite, hydromica, etc. as accessories. It is often distinctly schistose or foliated (**81–82, 85**), although usually massive, as in common soapstone (**89–90**). The foliated forms are generally the purest and may be pulverized for lubricating purposes; but the massive soapstone is, on account of its infusibility, toughness, and smoothness, admirably adapted to its important uses in the construction of stoves, sinks, aquaria, etc.

Pryophyllite Schist.—This is the massive or rock form of the aluminous talc — pyrophyllite. It is much like steatite (**93**) but rarer. Its principal uses are for slate pencils and tailor's chalk.

Pinite Schist.—This rather rare rock (**83**) is com-

monly to be regarded as an altered form of massive feldspathic rocks. Thus the pinite schist of Milton **(87)**
and other parts of the Boston Basin is simply a local al
teration of the felsites. Fragments of this rock are
common in the Roxbury pudding stone, although often
mistaken for serpentine.

Greensand.—This rock consists chiefly of the mineral glauconite, mingled usually with more or less sand,
clay, or calcareous matter. It is usually very friable, or
entirely unconsolidated. Although it is most abundant
in the newer geological formations, especially the Cretaceous **(88)** and Tertiary, it has a wide range in geological time. It is found in the Potsdam sandstone
(84) and all later formations; and is, perhaps, the only
one of the stratified silicate rocks now forming on an extensive scale in the ocean. Its value as a fertilizer, for
which purpose it is extensively employed, is due to the
potash which it contains.

Serpentine.—The rock forms of serpentine are presented under a great variety of conditions; and the
opinions of geologists with regard to its origin are
equally diversified. But, while it is very probable that
some forms of serpentine, especially those most intimately associated with limestone, are true sedimentary
deposits, it may now be regarded as certain that serpentine is chiefly an altered form of certain eruptive rocks,
such as basalt and peridotite. It is, therefore, only as a
matter of convenience, on account of the difficulty of
distinguishing the sedimentary from the eruptive serpentines, that they are all classed here with the metamorphic rocks. This classification is also partly justified by

the fact that the serpentines of all kinds are usually intimately associated with the metamorphic rocks—the gneisses, schists, marbles, etc. Most of the specimens exhibited are clearly altered eruptive rocks. The specimens from the Barc Hills (95-97), Hartz Mountains (82, 92), Elba (93) and Italy (89), may be mentioned as typical examples of this class. The specimen from the Hoosac Tunnel (91) represents one of the most important masses of serpentine in New England. Serpentine irregularly veined with magnesite is known as *verd antique* (81).

Chlorite Schist.—This rock stands in the same relation to the mineral chlorite that steatite or talc schist does to talc; and most of the specimens are so nearly pure chlorite as to be quite suitable for the mineral cabinet. Like steatite, the structure may be either massive, as in the very pure chlorite from Cross Island (85), or schistose (81-83).

It often contains more or less hydromica (83), and is sometimes distinctly talcose (86) as well as quartzose, feldspathic, and argillaceous. Chlorite schist and steatite are sometimes closely associated, as at Rowe, where they are worked in the same quarry (84), and the massive forms of chlorite schist have in a limited way the same uses as steatite. It is quite certain that chlorite schist, like serpentine, is sometimes, if not usually, an altered eruptive rock. This origin appears very probable for the foreign specimens especially (86).

(2) Eruptive or Unstratified Rocks.

The rocks of this great class are formed by the cooling and solidification of materials that have come up from a great depth in the earth in a melted and highly heated condition. When the fissures in the earth's crust reach down to the great reservoirs of liquid rock, and the latter wells up and overflows on the surface, forming a volcano, then we may, as has been explained on page 43, divide the eruptive mass into two parts: first, that which has actually flowed out on the surface, and cooled and solidified in contact with the air, forming a volcanic cone or sheet; second, that which has failed to reach the surface, but cooled and solidified in the fissure, forming a dike.

It has been shown that while the volcanic rocks or true lavas and the plutonic or dike rocks are essentially identical in composition, there is a marked difference in texture due to the widely different conditions under which the two classes have cooled and solidified. The plutonic or dike rocks solidify under enormous pressure, and this makes them heavy and solid — free from pores. They are, at the time of their formation, surrounded on all sides by warm rocks, which causes them to cool very slowly, and allows the various minerals time to crystallize.

The volcanic rock, on the other hand, cools under very slight pressure; and the steam, which exists abundantly in nearly all igneous rocks at the time of their eruption, is able to expand, forming innumerable small bubbles or steam-holes in the lava. Cooling in contact with the air, the lava cools quickly, and has but little chance for crystallization.   Hence, to summarize: plutonic rocks are solid and crystalline; and volcanic rocks are usually porous and uncrystalline.

Although the species and varieties of eruptive rocks which have been described are very numerous; yet it is sufficient for the purposes of general study and classification to recognize only four principal types in each of the two great divisions — the plutonic and the volcanic.   The feldspars and feldspathides are, with unimportant exceptions, essential and often the principal constituents of the various eruptive rocks, so that while the rocks of this great class show a general agreement in composition with the gneisses among the metamorphic rocks, they rarely exhibit any relationship in this respect with the schists. Again, the four principal types of the plutonic rocks are of strictly analogous and almost identical composition with those of the volcanic rocks.   Hence it is clear that in a general view we may consider that, so far as mere composition is concerned, and taking no account of their origins, the combinations of minerals observed in the gneisses are repeated under different conditions and textures among the plutonic rocks and again among the volcanic rocks.   The four principal feldspathic rocks in the metamorphic series, as we have seen, are gneiss, stratified syenite, stratified diorite, and norite.   The corresponding eruptive types are as follows: plutonic rocks — granite, syenite, diorite, and diabase; volcanic rocks — rhyolite, tractyte, andesite, and basalt.   The first two members of each series are acidic in composition, and the last two are basic.

The eruptive rocks are arranged after the manner of the metamorphic class; the plutonic division occupying the two upper shelves of sections 19-21, with the acid groups (granite and syenite) on the left, while the volcanic division occupies in the same manner the two lower shelves and the bottom of these sections.

Plutonic or Dike Rocks. — This division embraces, as already explained, all the eruptive or igneous rocks which, while in a liquid condition, have broken into but not entirely through the superficial portions of the earth's crust.   Hence they are also known properly as the irruptive rocks.   The plutonic rocks, having cooled slowly and under great pressure, are generally, like common granite, characterized by a dense and crystalline texture ; and, other things being equal, the coarseness of the texture is proportional to the depth below the surface at which the rock has solidified.   Hence, since the plutonic rocks can only be exposed as the result of erosion, which is a slow process, it is plain that, in general, the degree of crystallization affords a measure of the relative age of the rock and of the amount of erosion which the region has suffered since it was formed.

In the more elaborate lithological classifications we must recognize not only minor differences of composition and especially of alteration, but also, to some extent, differences of age, and the somewhat concomitant variations in the forms of the masses.   The regular wall-like dikes are especially characteristic of the relatively modern, superficial, and fine grained trappean rocks ; while the extremely irregular outlines are found chiefly with the older, deep-seated, and coarsely crystalline granitic rocks.

Granite. — Granite is the most abundant, the most varied and the most useful of all the eruptive rocks ; but the specimens on the first and second shelves of section 19 represent only a few of the numerous varieties.   As with gneiss, we properly begin with the simplest or binary form of granite, which consists of the essential minerals — quartz and feldspar — alone or without important accessories (**6–10**). The most finely granular varieties of binary granite are called aplite (**7–8**) or eurite (**10**). The beautiful Dedham granite is a good illustration of these. Many of the coarser red granites, such as those from the coast of Maine, New Brunswick, and Scotland (**12**) also belong here.

The specimens in the first row on this shelf represent the hornblendic variety (**1–5**), which was formerly called syenite, and is still often, but incorrectly, known by that name. The true syenite, as that term is now defined by lithologists, differs from granite in not containing any quartz. (See the next section.)

The original syenite (hornblendic granite) from the locality to which it owes its name — Syene, Egypt — is well exhibited in some of the ancient Egyptian sculptures in the Museum of Fine Arts, and in the Egyptian obelisk in Central Park, New York. Two of the hornblendic specimens (**2, 4**) illustrate the porphyritic texture in granitic rocks; while the specimens from Malacca and the Antarctic regions (**5, 11**) are interesting chiefly on account of the localities. It is to the hornblendic variety that most of the granite quarried at Quincy, Rockport, Peabody, etc., should be referred.

The specimens on the second shelf (**21–34**) belong chiefly to the micaceous variety, which is by far the most abundant and important form of granite. They show the usual range of textures, and the other features are sufficiently explained on the labels. The drawing (**3**) shows the appearance of a thin section of granite when magnified in polarized light; and attention is called particularly to the liquid inclusions in the brightly colored grains of quartz, and to the fact that the extreme irregularity of the forms of the quartz proves that it was the last of the principal constituents to crystallize.

Syenite.— As compared with granite, the true syenite or quartzless granite is a rare and economically unimportant rock. The majority of the specimens (1-6, 26, section 20) represent the principal variety, which is composed of the essential mineral — feldspar, chiefly red and gray orthoclase, and hornblende. The hornblende is not infrequently replaced partly or wholly by black mica (3) ; and the orthoclase in part by plagioclase or by elaeolite, one of the feldspathides (4). Zircon is one of the most interesting accessories, forming the variety zircon syenite (5). *Greisen* (6) is a massive aggregate of quartz and mica, and may be referred to in this connection, although its eruptive origin is very doubtful.

Diorite. — This is one of the most abundant rocks in Eastern Massachusetts; it presents, however, but few distinct or interesting varieties and it is of no economic importance. For the most part it is a finely crystalline and monotonous aggregate of plagioclase and hornblende with more or less black mica and epidote (11-16, 31-37) ; but one of the specimens (14) shows that more interesting accessories sometimes occur. *Miascite* (32) and *kersantite* (38) are rare species belonging in this part of the classification.

Diabase. — Broadly but correctly defined, diabase is the most important and interesting of the basic plutonic rocks. It is the principal dike-forming rock of this region. Nearly all the more or less regular or wall-like "trap" dikes cutting through the slate, conglomerate, granite, etc., about Boston and along the coast, consist of diabase and it is, therefore, a rock of especial interest for students of our local geology. Although

normally it is a crystalline granular aggregate of plagioclase, augite, and magnetite, it presents many varieties of
texture and composition.   It may be very coarsely crystalline or porphyritic, or quite compact; the compact or
aphanitic forms being the typical trap which occurs so
abundantly in the smaller dikes (1, section 21).   The
porphyritic diabase (2, 4-5) is sometimes of interest as
an ornamental stone; and some of the antique porphyry
(10) belongs here.   Chrysolite is not an uncommon accessory mineral (9) in diabase, and it is interesting especially as making the identity of composition of diabase
and basalt more complete.   The augite, especially in the
older and more coarsely crystalline diabase, is often
partly or wholly replaced by diallage or hypersthene,
giving the varieties or sub-species gabbro (24-29) and
hyperite (25).   Greenstone, although a much abused
lithological term, may now be most properly used to
designate the important variety of diabase in which the
original minerals, and especially the augite, have been
largely changed to chlorite or similar green hydrous silicates (21, 26).   The serpentinic diabase (36-37) may
be regarded as a special phase of greenstone.
    The drawing (3) represents a thin section of typical
diabase magnified in polarized light, and shows especially the slender twinned crystals of plagioclase.
Peridotite and *dunite* or *olivine rock* (31-35) may be
classed with the plutonic rocks, although it is probable
that they belong partly in the volcanic division.   Peridotite is essentially a granular aggregate of chrysolite
or olivine, with often more or less plagioclase, augite, magnetite, chromite, etc.   In fact it may be usu-

ally regarded as a very highly chrysolitic diabase; *i.e.*, as a diabase in which this accessory has become the chief essential constituent.

Volcanic Rocks or Lavas. — The general characters of the true lavas or surface flows have been noted. It may be added here that, while composed of essentially the same minerals as the plutonic rocks, there are some notable differences, such as the general absence of white mica; and that although composed in part at least, in every typical and unaltered example, of amorphous matter or glass, three more or less distinct varieties based upon texture may be recognized. These textural phases are: (1) *holocrystalline*, when the rock is composed chiefly of visible crystals, as in rhyolite, trachyte, andesite, and basalt; (2) *vitreous*, when the volcanic glass predominates, as in obsidian and tachylite; and (3) *devitrified or felsitic*, when an original vitreous texture has been changed to a stony or compact texture, as in petrcsilex, felsite, porphyrite, and melaphyr. They may also be designated as the *crystalline*, *uncrystalline* or *amorphous*, and *semi-crystalline* forms of volcanic rocks. The first is illustrated by the specimens on the third shelf in sections 19-21; the second by those on the fourth shelf; while the devitrified or semi-crystalline rocks occupy the bottom of these sections. It is most convenient and natural, however, to treat this textural classification as subordinate to the acidic and basic groups; and in the following paragraphs each of the latter will be described as a whole.

Rhyolite and Trachyte. — The trachytic rocks, including rhyolite (**41-57**, section 19), which corresp onds

in composition with granite, but is less commonly micaceous, and trachyte proper (**41–50**, section 20), which
has very exactly the composition of syenite, are usually
imperfectly crystalline to compact rocks, characterized by
a somewhat porous texture and harsh or rough feel.   All
these features, as well as the characteristic light colors,
are well illustrated by the specimens, which represent
some of the more important regions in which trachytic
rocks occur.   The chief difference between rhyolite and
trachyte is that the former contains quartz ; but the minerals are often so imperfectly developed as to make it
difficult to determine whether quartz is present or not.
It shows distinctly in only a few of the specimens on the
shelf (**48–49**).   Sanidin, the clear form of orthoclase,
found only in volcanic rocks, is well shown in several
specimens of both rhyolite (**47**) and trachyte (**42**), being
porphyritically developed in large crystals.

Obsidian.—Obsidian is a true volcanic glass and may
be defined as either rhyolite or trachyte, chiefly the
former, which has cooled so rapidly as to remain wholly
or partially in an amorphous or uncrystalline state, being sharply distinguished from all other rocks by its
perfect vitreous texture.   The specimens illustrate the
principal varieties, which are based chiefly upon the
secondary textures.   The plain, black glass (**61–62, 66–
67,** section 19) may be regarded as the most acid and the
most typical obsidian, the gray and relatively opaque
specimens (**65, 70**) probably agreeing more closely in
composition with trachyte than with rhyolite.   No volcanic rock affords finer illustrations of fluidal lines or
flow-structure than the banded obsidians (**64, 68**).

The breccia texture (69) naturally results from the mass continuing to flow after the superficial portions of it have become solid. The vesicular texture (61, 63, 65, section 20) testifies with equal distinctness to the former fluid state of the rock ; the vesicles or steam-holes being due to the expansion of the contained vapors when the rock was still molten. The most highly and finely vesicular obsidian is the typical pumice (61). The flowing of the frothy mass is often indicated by the elongation of the steam-holes, a more or less distinctly fibrous structure being developed in this way.

Although the typical obsidian is a true glass, and entirely amorphous, a more or less marked tendency to crystallization may be exhibited in several ways. First, in the development of microscopic crystals or crystallites, which may be so numerous as to amount to a partial devitrification of the obsidian, as in the variety *pitchstone* (73, 75). Second, in the aggregation of the crystallites in concretionary masses or spherules, as in the variety *spherulite* or spherulitic obsidian (70, section 19, and 64, section 20). Third, in the development of true, macroscopic crystals, as in porphyritic obsidian (72, 76). *Perlite* (62, section 20) is an interesting structural variety due to local contraction and concentric splitting during cooling.

Petrosilex and Felsite.—The felsites or felsitic rocks include petrosilex, which agrees in composition with rhyolite and the more acid and typical obsidian, and felsite proper, which shows a similar agreement with trachtye and the less acid obsidian. In other words, the felsites, as previously explained, are either acid lavas which have cooled too slowly to permit the formation of a true glass (obsidian) and yet too rapidly

for the development of a distinct crystalline texture
(rhyolite or trachyte) ; or they are altered obsidian, *i.e.*,
obsidian which has suffered devitrification subsequently
to its original cooling. The intimate genetic relation of
obsidian and felsite is seen in the fact that nearly every
variety or structural feature of the former is observed
in the latter. The felsites, have, like the granites and
traps, a large and varied development in the region
about Boston, ranking as, perhaps, the most interesting
and beautiful feature of our local geology. The most
typical petrosilex and felsite exhibit the semi-crystalline
or felsitic texture throughout, as in the so-called
"Saugus jasper" (**81-86**),which may be compared with
the plain glassy obsidian. Every phase of the flow-
structure of obsidian is most perfectly reproduced in the
banded petrosilex (**83, 87-88**) and felsites (**90, 98**) ; and
the same is substantially true of the brecciated (**95-97**),
concretionary or spherulitic (**99**), and porphyritic (**84-
85**), textures. The felsites, however, are much more
commonly porphyritic than obsidian, and unlike obsidian
are often porphyritic with quartz as well as feldspar,
giving the variety quartz porphyry (**85-86**, section 20),
which frequently exhibits a gradual passage into granite
(**92-93**). The porphyritic texture is so characteristic of
these rocks that they are commonly known as porphy-
ries: and among our native varieties there are some
which appear to be as suitable as the antique porphyry
(**81-84**, section 20), for architectural and ornamental
uses.

　　Andesite and Basalt.—These are the basaltic, as
distinguished from the trachytic, rocks, and have, as the

labels show, approximately the composition, respectively, of diorite and diabase. A glance at the specimens shows that they are much darker colored than the trachytes, although generally similar to them in crystallization and texture. The andesites (section 20) differ chiefly in the proportions and nature of the hornblendic and feldspathic constituents, ranging from highly feldspathic varieties (**51, 53**) to those which are black with hornblende and magnetite (**52**). The presence of quartz gives the variety *dacite* (**59**).

Basalt (section 21) is more varied. As regards texture it may be crystalline, compact (**47**), or vesicular (**58**). Chrysolite, in clear glassy green grains, is the chief accessory mineral (**45,48**). In the lavas of Mt. Vesuvius and some other volcanoes the plagioclase is partly replaced by leucite in gray isometric crystals (**50, 53**). Nephelite and other feldspathides also play the same role. The basalt from Lassen's Peak (**52**) is interesting on account of containing quartz as a prominent constituent. Many lavas have an historical interest. This is especially true of the eruption from Mt. Vesuvius in the year 79 (**59**), the first recorded eruption of that crater, and the one which overwhelmed the cities of Pompeii and Herculaneum.

Tachylite.—This species includes the more or less distinctly amorphous or glassy forms of both andesite and basalt, standing in the same general relation to these rocks that obsidian does to rhyolite and trachyte. It is usually, however, a much less perfect glass than obsidian, the basic crystallizing more rapidly and perfectly than the acidic lavas. The most glassy tachylite is quite.

opaque, resembling certain black furnace slags (62, 64, section 21). The flowing of the liquid lava is commonly indicated by very characteristic surface features (61, section 21, 71, section 20). Tachylite is very generally vesicular or scoriaceous, as seen in most of the specimens, although rarely so thoroughly vesicular and spongy as obsidian in the form of pumice. The almost perfect liquidity of the molten lava, even when rapidly cooling, is beautifully illustrated by the stalactitic forms (69, 73–74) resulting when the lava falls, drop by drop, from some overhanging surface. Pele's hair (65) is the interesting variety produced when the highly liquid lava is caught up and blown by the wind, a veritable frozen spray.

As in basalt, the most important accessory mineral is chrysolite (74-75). Several of the specimens represent flows whose dates are known; and some of these are quite recent, as the great eruption of Mauna Loa in 1886 (74-75), and the small eruption of Etna in 1883 (66). Vesuvius is now almost continuously active, and the specimens of fresh lava thrown out in 1883 (67) might be duplicated at almost any time. Some of the older dated specimens are interesting at least for the great magnitude of the eruptions which they represent (75, section 20). Although the fresh basaltic lavas are naturally dark colored or black, they commonly weather reddish, through the peroxidation of the iron (74, section 20).

Porphyrite and Melaphyr.—These may be regarded as semi-crystalline basaltic lavas, or devitrified forms of tachylite ; holding very much the same relations to this rock that petrosilex and felsite do to obsidian. Comparatively modern and unaltered forms of each are well illustrated by the dark specimens from Germany (**91,** section 20, **81,** section 21). In a highly altered condition, in which the original augite, feldspar, and olivine have been largely changed to epidote, chlorite, quartz, etc., they are extremely abundant rocks in the vicinity of Boston, including the most of the basic lavas of the Boston Basin. These ancient lavas exhibit nearly every structural feature observed in the products of modern craters, except that the steam-holes or vesicles have been very generally filled up by the secondary minerals — quartz, epidote, etc., changing the amygdaloidal to the vesicular texture (**84-85**). They are very commonly brecciated, after the manner of recent lavas (**86**); and in some cases the characteristic surface contours have been preserved.

Vein Rocks.

All rocks are not embraced in the sedimentary and eruptive divisions, but there is a third grand division, which it is deemed best to include in the illustration. These are the vein rocks. They present a large number of varieties, and yet, taken altogether, form but a small fraction of the earth's crust. They are, however, the great repositories of the precious and other metals, and hence are objects of greater importance to the miner and practical man than the eruptive rocks, or, in some parts of the world, even than the sedimentary rocks. The specimens illustrating this division of lithology occupy sections 22 and 23, between the windows.

The vein rocks, like the plutonic rocks, occupy fissures in the earth's crust intersecting the stratified formations; but the fissures filled with vein rocks are called veins and not dikes. The mode of formation of a typical vein has been explained on page 48.

The water circulating through the earth's crust is often saturated with its various mineral constituents, and veins are formed by the deposition of the dissolved minerals in fissures. One of the most important characteristics of the vein rocks, as a class, is the great variety which they present; for nearly every known mineral is embraced among their constituents; and these are combined in all possible ways and proportions, so that the number of combinations is almost endless. The solvent power of the subterranean waters varies for different minerals; and appears often to be greatest for the rare species. In other words, there is a sort of selective action, whereby many minerals which exist in stratified and eruptive rocks so thinly diffused as to entirely escape the most refined observation are concentrated in veins in masses of sensible size; and our lists of known minerals and chemical elements are undoubtedly much longer than they would be if these wonderful storehouses of fine minerals which we call veins had never been explored. As a rule, the minerals in veins form larger and more perfect crystals than we find in either of the other great classes of rocks. This is simply because the conditions are more favorable for crystallization in veins than in dikes or sedimentary strata. In both dikes and strata, the growing crystals are surrounded on all sides by solid or semi-solid matter; and, being thus hampered, it is simply impossible that they should become either large or perfect. In the vein, on the other hand, there are usually no such obstacles to be overcome ; but the crystals, starting from the walls of the fissure, grow toward its center, their growing ends projecting into a free space, where they have freedom to develop their normal forms and to attain a size limited only, in many cases, by the breadth of the fissure. With, possibly, some rare exceptions, all the large and perfect crystals of quartz, feldspar, mica, beryl, apatite, fluorite, and of minerals generally, which we see in mineralogical cabinets, have originated in veins. Those fissures which become the seats of mineral veins are really Nature's laboratories for the production of rare and beautiful mineral specimens ; and hence the vein rocks are the chief resort of the mineralogist, to whom they are of far greater interest than all the eruptive and stratified rocks combined.

The leading characteristics, then, of the vein rocks may be summarized as follows: (1) They contain nearly all known minerals, including many rare species and elements which are unknown outside of this class of rocks. (2) These mineral constituents, occurring singly and in association, give rise to a number of varieties of rocks so great as to baffle detailed description. (3) They exceed all other rocks in the coarseness of their crystallization, and in the perfection and beauty of the single crystals which they afford.

The arrangement, and to a large extent the nomenclature, of the specimens is essentially mineralogical, beginning with the sulphides and ending with the silicates. They have been selected with a view to showing a few only of the more typical and interesting minerals and associations of minerals occurring in veins. Some show complete sections of small veins (gypsum) ; and the large geode may be regarded as an incomplete vein. Others exhibit characteristic vein structures, with prismatic crystals perpendicular to the walls (graphic granite and quartz). The columnar and stalactitic forms (limonite and calcite) manifest the same general tendencies.

These two sections may be regarded as one, the specimens and numbers continuing across on each shelf from 22 to 23. The first specimens are examples of the metallic ores, which occur very largely in veins. The galenite (1) represents the peculiar and important lead veins in the limestones of the Mississippi Valley. The zinc ore (3) and copper ore (4) are from veins of more normal type in the crystalline rocks of the East. Pyrite (5) is the most important mineral in many of the auriferous veins of the Rocky Mountains; the more massive specimen (6), however, represents the great vein in Rowe, Mass., which is mined to get pyrite for the manufacture of sulphuric acid. The next specimens (11–12) are additional examples of copper and zinc ores, which show by their forms that they must have come from veins. The iron ores are mainly, like the coals and bitumens, bedded rocks, but these examples of specular hematite (13–14), magnetite (67), and limonite (81) are quite clearly the products of veins, and albertite (15), a variety of asphaltum, is known to fill fissures cutting across the strata. The large mass of native copper (91) is from a vein and illustrates one important mode of occurrence of the copper in the Lake Superior mines.

Calcite, although occurring chiefly in the form of a stratified rock (limestone), is also an important vein-forming mineral. It is often found coarsely and beautifully crystallized (26, 36-37) ; but it is still more commonly massive or quite compact, as in the cavern deposits—stalactites (32) and stalagmites (31). This is essentially the origin of the beautiful onyx marbles (21-22). What has been stated for calcite might be repeated for gypsum (22-24, 28), and barite (33, 35). Apatite, in its mineral forms (25-27), occurs chiefly in veins, as the large and perfect crystals indicate. Fluorite (43) is another important vein-forming mineral, being, like calcite, barite, and quartz, a common constituent of metalliferous veins, the gangue or matrix of the ores. The vein forms of the most important of all the vein-forming minerals—quartz—are illustrated in a general way by the specimens (41-42, 44-45), the massive, vitreous quartz (45) being the most abundant. The large geode (82) may be regarded as a half-formed globular vein of chalcedony and crystalline quartz.

Among the ancient crystalline formations the veins consist very largely of either quartz, or quartz and various silicates, such as the feldspars, micas, etc. Of this character are the great veins of coarse or giant granite which in various parts of New England are quarried for commercial quartz (45), feldspar (53-56), and mica (74-75); and for crystallized cabinet minerals, such as tourmaline (71), beryl (73), etc. The large group of beryls in quartz in the Vestibule shows with what a lavish hand Nature has furnished these mineralogical store houses, while the enormous single crystal of beryl shows how favorable the conditions have been for the development of the mineral individuals. Graphic granite or pegmatite (62-64, 72) is a very characteristic structural feature of the less coarsely crystalline portions of the vein granite. The large specimen from Fitchburg (65) shows a more nearly complete section of a granite vein; and the highly micaceous specimen from the Black Hills (61) represents the great tin-bearing vein in which crystals of spodumene thirty to fifty feet in length and one to three feet in diameter have been observed.

PETROLOGY.

In lithology we investigate the nature of the materials composing the earth's crust — the various minerals, and aggregates of minerals, or rocks ; while in petrology we consider the forms and modes of arrangement of the rock-masses, — in other words, the architecture of the earth.

Petrology is the complement of lithology, and in many respects it is the most fascinating division of geology, since in no other direction in this science are we brought constantly into such intimate relations with the beautiful and sublime in nature. The structures of rocks are the basis of nearly all natural scenery ; for what we call scenery is usually merely the external expression, as developed by the powerful but délicate sculpture of the agents of erosion — rain and frost, rivers and glaciers, etc., — of the geological structure of the country. And to the practised eye of the geologist, a fine landscape is not simply a pleasantly or grandly diversified *surface*, but it has *depth*; for he reads in the superficial lineaments the structure of the rocks out of which they are carved.

But, while the magnitude of the phenomena adds greatly to the charm of the study, it also increases the difficulty of procuring suitable illustrations for the museum and class room. Nature, however, has not been wholly unmindful of our needs in this direction ; for she has worked often upon a very small as well as a very large scale; many of the grandest phenomena being repeated in miniature. Thus, we observe rock-folds or arches miles in breadth and forming mountain masses, and of all sizes from that down to the minutest wrinkle. So with veins, faults, etc. ; and the wonderful thing is that these small examples, which may be placed in the cabinet, are usually, except in size, exactly like the large. Now the aim in this department of the Museum has been to secure as complete a collection as possible of these natural models; and to supplement these to only a limited extent by artificial illustrations. It is practically impossible to illustrate all the topics of petrology, without adding greatly to the artificial character of the collection by the free use of pictures and diagrams, and thus encroaching upon the proper ground of the text-book. And since the Guide is necessarily limited to the description and explanation of the objects in the collection, those desiring a more comprehensive and systematic treatment of petrology are referred to the standard text-books of the science, and to No. XII of the series of Science Guides published under the auspices of the Society.

The Petrological Collection, to which this section of the Guide relates, is contained in the two central or floor cases in Room B.

CLASSIFICATION OF STRUCTURES.

The structures of rocks divide at the outset into two classes: (1) the *original structures*, or those produced at the same time and by the same forces as the rocks themselves, and which are, therefore, peculiar to the class of rocks in which they occur (stratification, ripple-marks, fossils, etc.); and (2) the *subsequent structures*, or those developed in rocks subsequently to their formation, by forces that act more or less uniformly upon all classes of rocks, and which are, therefore, in a large degree, common to all kinds of rocks (folds, faults, joints, etc.).

The original structures are conveniently and naturally classified in accordance with the three great divisions of rocks: (1) stratified rocks, (2) eruptive rocks, and (3) vein rocks; while the subsequent structures, not being peculiar to particular classes of rocks, are properly divided into those produced by: (1) the subterranean or igneous agencies, and (2) the superficial or aqueous agencies. The original structures, with which we begin, are illustrated by the specimens in the first case (sections 1 to 8), and the subsequent structures by those in the second case (sections 9 to 16).

ORIGINAL STRUCTURES OF STRATIFIED ROCKS.

Stratification. — All rocks formed by strewing materials in water, and their deposition in successive, parallel, horizontal layers, are *stratified* ; and stratification is not only the chief structure resulting from this process, but it is the most important of all rock-structures. The first specimens in this section (1) are good examples of distinct and regular stratification in different kinds of rocks : iron ore (**1**), mica schist (**2**), gneiss (**3-4**), sandstone (**21-23, 41**), bituminous coal (**42**), clay (**43**), and, on the fifth or bottom shelf, slate (**87**). The stratification is, however, often not apparent to the eye ; as in most of the specimens on the third shelf. It shows indistinctly in the granitoid gneiss (**48**), but is wholly wanting in the marble (**47**), cannel coal (**49**), freestone (**50**), and chalk (**51**). The explanation is easily found, for any one can readily prove by an experiment with clay or fine sand in a vessel of water that if precisely the same kind of material is deposited continuously and uniformly there will be no visible stratification in the deposit, because there is nothing in the nature of the sediment or the way in which it is laid down to develop distinct lines of stratification. Continuous and uniform deposition obtains very frequently in nature, as the specimens indicate ; but it rarely continues long enough to permit the formation of thick beds or strata. Hence, while the stratification is almost always visible on the large surfaces of sandstone, slate, etc., exposed in quarries and railway cuttings, and may usually be seen in the quarried blocks, it is often not apparent in smaller masses or hand specimens, which may represent a single homogeneous layer. There is one important exception, and that is where the particles, although of the same kind, are flat or elongated. Pebbles of these forms (61) are common on many beaches ; and since they are necessarily arranged horizontally by the action of the water, they will, by their parallelism, make the stratification of the resulting pudding-stone (65) visible. The same result is accomplished still more distinctly by the mica scales in many sandstones (62), the leaves and flattened stems of vegetation in bituminous coal (42), shale (63), and the flat shells in limestone (66).

Changes in a rock subsequently to its deposition will also sometimes develop or bring out the stratification where it was before invisible. Thus, the cherty limestone (67) was originally quite homogeneous and massive, but the segregation of the disseminated silica to form the layers of chert makes the stratification very apparent.

In all other cases, visible stratification implies some change in the conditions ; either the deposition was interrupted or different kinds of sediment were deposited at different times. The first cause produces planes of easy splitting, or fissility, especially in fine-grained rocks, like shale (64). This shaly structure or lamination-cleavage may be due, in some cases, to pressure, but it is commonly understood to mean that each thin layer of clay became partially consolidated before the next one was deposited upon it, so that the two could not perfectly adhere; although the imperfect adhesion of the layers must be attributed in many cases to the deposition between them, at times when the water is unusually quiet, of slight films of finely divided mica, organic detritus, etc. These parallel planes of easy splitting are, however, by themselves, of little value as indications of stratification, since the lamination-cleavage is not always easily distinguished from true slaty cleavage (roofing slate), and parallel jointing structures developed subsequently to the deposition of the sediments and quite independent of the stratification. The second cause, or variations in the kind of sediment, gives alternating layers differing in color, texture, and composition, as is seen in most of the specimens of sandstone, slate, coal, iron ore, gneiss, etc. already referred to; and of all the indications of stratification these are the most important and reliable.

These specimens also illustrate well the division of stratified rocks into strata and laminae. A layer composed throughout of essentially the same kind of rock, as conglomerate or sandstone and showing no marked planes of division, is usually regarded as one *stratum* or *bed*; while the thinner portions composing the stratum and differing slightly in color, texture, or composition, and the thin sheets into which the shaly rocks split, are the *laminae* or *leaves*. The geological record is written chiefly in the sedimentary rocks; and the formations, strata, and laminae may be regarded as the volumes, chapters, and pages in the history of the earth. It is especially important to note in this connection that every line of stratification and every change in the

character of the sediments is due to some change of corresponding magnitude in the conditions under which the rock was
formed. The slight and local changes in the conditions occur
frequently and mark off the individual laminae and strata, while
the more important and wide-spread changes determine the
boundaries of the groups of strata and the geological formations.

Strata are subject to constant lateral changes in texture and composition, *i.e* , a bed or formation rarely
holds the same lithological characteristics over an extended area. There are some striking exceptions, especially among the finer-grained rocks, like slate, limestone,
and coal, which have been deposited under uniform conditions over wide areas. It is the general rule, however, particularly with the coarse-grained rocks, which
have been deposited in shallow water, near the land,
that the same continuous stratum undergoes great
changes in thickness and lithological character when followed horizontally. A stratum of conglomerate becomes
finer grained and gradually changes into sandstone,
which shades off imperceptibly into slate, and slate into
limestone, etc. Where the stratum is conglomerate, its
thickness will usually be much greater and more variable
than where it is composed of the finer sediments. The
rapidity of these changes in certain cases is well shown
by the parallel sections in the drawing (**44**). These
represent precisely the same beds, as the connecting
lines indicate, at points only twenty feet apart.

When we glance at the conditions under which stratified
rocks are now being formed, it is plain that all strata must ter-
minate at the margin of the sea in which they were deposited,

and in the marginal portions of that sea, especially, must exhibit frequent and rapid changes in composition, etc. The sediments forming the surface of the sea-bottom at the present time may be regarded as belonging to one continuous stratum; and it is instructive to examine a chart of any part of our coast, such as Massachusetts Bay, on which the nature of the bottom is indicated for each sounding, and observe the distribution of the different kinds of sediment. On an irregular coast like this, especially, the gravel, sand, and mud of different colors and textures, and the different kinds of shelly bottom, form a patchwork, the patches being, for the most part, of limited extent and shading off gradually into each other.

On a more regular coast, like that of New Jersey, the sediments are distributed with corresponding uniformity, the changes are less frequent and more gradual; and we have here a better chance to observe the normal arrangement of the sediments along a line from the shore seaward—gravel, sand, mud, and shells. On the beach we find the shingle and coarse pebbles, shading off rapidly into fine pebbles and sand. The zone or belt of sandy bottom may vary in width from a mile or two to twenty miles or more, becoming gradually finer and changing into clay or mud, which covers, usually, a much broader zone, sometimes extending into the deeper parts of the sea, but gradually giving way to calcareous sediments. Hence we may say that the finer the sediment the greater the area over which it is spread; but, on the other hand, the coarser the sediment the more rapidly it increases in thickness. In other words, the horizontal extent of a formation deposited in any given period of time is inversely, and the vertical extent or thickness is directly, proportional to the size of the particles.

Overlap and Interposition of Strata. — It has been shown on pages 31 to 36 that nearly all parts of the earth's surface, including both the land and the sea-floor, are now and probably always have been either rising or sinking; and these slow movements of

elevation and subsidence, which must at long intervals be
reversed in direction, involve a corresponding retreat or
advance of the shore-lines of the ocean. One important
consequence of these constant oscillations of the shore-
lines is that successive deposits in the same sea will often
cover different and unequal areas. When, in consequence
of subsidence, one formation extends beyond and covers
the edge of another, as shown in the drawing (**45**), we ·
have the phenomenon described as *overlap*. *Interposition*
is similar, being the case where a formation (**46**) does
not, in certain directions, cover so wide an area · as the
strata above and below it, which are thus sometimes
found in contact, although normally separated by the entire thickness of the intervening and, seemingly, interposed stratum.

Unconformity.—The model (**86**) is a good general
illustration of this important feature of stratified rocks ;
and the label explains in detail how the successive uncon-
formities of the strata prove that the area in question has
been repeatedly elevated and depressed. We have
already learned (pages 52 to 69) that the rocks on
the land are being constantly worn away by the
agents of erosion ; and it is a matter of common observation that the strata thus exposed are often not horizontal, but highly inclined, having been greatly disturbed
and compressed during their elevation to form dry land.
Now, when such a land-surface subsides to form the sea-
bottom, and new strata are spread horizontally over it,
they will lie across the upturned and eroded edges of the
older rocks, and fill the hollows worn out of the latter,
as shown in this model ; and the new formation is then

said to rest unconformably upon the older. Observation shows, however, that strata are often elevated to form dry land and exposed to erosion while still retaining their horizontal position. Hence, as the drawings indicate, we must broaden the definition of unconformity, as follows : Two strata or formations are unconformable when the older has suffered erosion (**85**), or disturbance (**84**), or both disturbance and erosion (**83**), before the deposition of the newer.

When strata are conformable, the deposition may be presumed to have been nearly or quite continuous ; but unconformity clearly proves a prolonged interruption of the deposition during which the elevation, erosion, and subsidence of the sea-bottom took place. The specimen of conglomerate from Nantasket (**89**) shows how an unconformity may be proved when the actual contact of the two formations can not be seen, if the newer formation is a conglomerate containing fragments of the older. In this instance the pebbles or water-worn fragments of granite in the conglomerate, which must have been worn from solid ledges, show that the granite was exposed above the level of the sea and suffered erosion before the formation of the' conglomerate. By this criterion we know that the conglomerate of the Boston Basin (the Roxbury pudding-stone) rests unconformably not only upon the granite, but also upon the quartzite, felsite and other ancient rocks of this region.

Irregularities of Stratification.—These are especially noticeable in the gravel and sand rocks, or those which have been deposited chiefly by strong, local, and variable currents ; the kind and quantity of sediment, of course, varying with the strength and direction of the currents. The numerous sand and gravel banks about Boston afford admirable illustrations of nearly every phase of irregularity; and only two kinds require special illustration.

(1) *Contemporaneous erosion and deposition*, where, in consequence of a change in the currents, fine material, like sand, recently deposited, is partially swept away and its place taken by coarser sediments, such as gravel. This structure can scarcely be illustrated by a cabinet specimen; but the drawing (**82**) shows substantially what can be seen in many of our gravel banks.

(2) *Oblique lamination or current-bedding*, where the strata may be horizontal, as usual, but their component laminae are inclined at various angles. The specimen of quartzite (**68**) is an excellent illustration, showing a single thin bed, with the laminae oblique, instead of parallel, to its upper and lower surfaces. This structure is characteristic of sediments swept along by strong currents and deposited in shallow basins or depressions of the sea-bottom or of a river-bed. On a larger scale, also, it is the normal structure of deltas. Many of the level plains of sand and gravel in this vicinity are delta deposits formed in temporary lakes; and this is proved by the fact that they often afford, where recent excavations have been made, sections like that represented in the drawing (**81**). The current or stream has moved in the direction indicated by the arrow, and deposited its load of sediment in the quiet waters of an ancient lake as shown in the companion drawing on this tablet. The successive increments of fine material are carried to the growing edge

of the delta and deposited as sloping layers (*foresets*), as-
suming the maximum angle of stability of the material;
while the coarser sediment, the coarsest sand and gravel,
is arrested before reaching the outer edge of the delta,
and forms the horizontal layers (*topsets*) seen in the upper part of the section.

Ripple-marks.—This interesting structure is illustrated by an exceptionally good series of specimens.   All
who have been on a beach or sand bar must have noticed
the lines of wavy ridges and hollows, or ripples, on the
surface of the sand.   These are sand waves, produced
by water moving over the sand, or by air moving over
dry sand, very much as ordinary waves are formed by
air moving over water.   Each tide usually effaces the
ripple-marks made by its predecessor and leaves a new
series, to be obliterated by the next tide.   But where
sediment is constantly accumulating, a rippled surface
may be gently overspread by a new layer, and thus preserved.   Other series of ripples may, in like manner, be
formed and preserved in overlying layers, as seen in the
photograph (**2**) of a sand-bank near Mt. Hope Station;
and when the beach becomes a firm sandstone, a section
of it will show the rippled surfaces almost as perfectly as
when they were first formed, as may be seen in several
of the specimens.

Ripple-marks are most perfect in fine sand or sandstone, as the specimens show.   They are not formed in
gravel, because it is too coarse; and rarely in clay, because it is too tenacious, and is deposited where the water
is too quiet.   Two of the specimens (**21–22**) show, however, that ripples in argillaceous sediments may be more

sharply defined, as well as more prominent in proportion
to their size, than those formed in sand.

The specimen of rippled limestone (**1**) may be regarded
as very exceptional in the nature of the material. Ripple-
marks are usually limited to shallow water, or at least to
places where strong currents or waves sweep over sandy
bottoms : and hence are regarded as proving that the
rocks in which they occur are shallow-water or beach deposits. They are normally at right angles to the current
that produces them, and where this changes with the
direction of the wind, cross-ripples and other irregularities are introduced. Ripple-marks are also usually parallel with the beach ; and when they are found on the
ledges they give us the direction, as well as the position
of the ancient shore-line.

Again, the friction of the water pushes the sand-grains
along, rolling them up on one side of the ripple and let-
ting them fall down on the other. Hence, ripples being
formed by a current are always moving in the same direction as the current, and are usually unsymmetrical on
the cross-section ; presenting a long, gentle slope toward
the current, and a short, steep slope away from it. The
specimen of brown sandstone from Wyoming (**42**) is a
very good illustration of the current-ripples ; and it is
evident that when we observe these fossil ripples in the
original ledges instead of in detached specimens we may
learn from them, not only the position and direction of
the ancient shore, but also, in some cases, on which side
the land lay and on which side the sea.

When the water is in a state of oscillation, without any
distinct current, more symmetrical ripples are produced,

such as are seen in most of the specimens, especially in the large slabs from the Connecticut Valley (**4, 24**), Port Henry (**42**), and Brighton (**88**, section 1). In one of the Brighton specimens (**44**) the ripples are exceptionally large, indicating quite a violent oscillation of the water.

Ripples, and especially those due to oscillation of the water, often present somewhat sharp and angular crests and smoothly rounded hollows or troughs. Now the sand or mud which covers a rippled surface, forms, when it is hardened, a cast of the ripples; and the characters just explained—sharp crests and rounded troughs, enable us to determine whether a specimen shows the original rippled surface or a natural cast of it. By this test we should judge that some of the examples already referred to (**1, 43**) are not genuine ripples, but casts. One of the Brighton specimens (**3**) is of particular interest in this connection, since it shows both phases and permits a direct comparison. It has been split along a rippled surface, and the front piece shows the actual ripples and the back piece the casts. This principle has been made use of where the rocks are greatly disturbed, to prove an actual inversion of the strata, the ripples being upside down. The large slab of sandstone from Turner's Falls, in the passage between Room A and Room B, is a good example of current ripples; and ripple-marks are shown less distinctly on the larger slabs in the Vestibule.

Rill-marks, Rain-prints, Sun-cracks, etc.—"One of the most fascinating parts of the work of a field-geologist consists in tracing the shores of former seas and lakes, and thus reconstructing the geography of successive geological periods." His conclusions, as we have already seen, are based largely upon the nature of the sediments; but still more convincing is the evidence afforded by those superficial features of the strata, which, like ripple-marks, seem, by themselves, quite insignificant. And among these he lays special emphasis upon those which show that during their deposition strata have at intervals been laid bare to sun and air.

During ebb-tide, water which has been left at the upper edge of the beach runs down across the beach in small rills, which excavate miniature channels or rill-marks; and when these are preserved in the hard rocks, they prove that the latter are beach deposits, and, like the ripple-marks, show the direction of the old shore. The specimens from Pennsylvania (**61**) are well preserved fossil rill-marks, and essentially similar to what may be often seen on the beaches of the present day.

If a shower of rain falls on a muddy beach or flat, the sediment deposited by the returning tide may cover, without obliterating, the small but characteristic impressions of the individual drops; and these markings or rain-prints are frequently found well preserved in the hardest slates and sandstones, testifying unequivocally to the conditions under which the rocks were formed. The illustrative specimens are all from the Triassic sandstones and shales of the Connecticut Valley; and they prove that this valley was once a broad and shallow estuary, or arm of the sea. On two of the specimens (**63–64**) the prints are quite distinct and perfect, while the others (**81–82**), and the large slabs in the Vestibule, show them less distinctly and may be supposed to represent or record a prolonged shower or settled rain, the earlier impressions being obscured or obliterated by the later ones. In some cases the rain-prints are found to be ridged upon

‚one side only, in such a manner as to indicate that as the
drops fell they were driven aslant by the wind. The
prominent side of the marking, therefore, indicates the
side toward which the wind blew. The fossil rain-prints
should be compared with the artificial or recent specimens on the upper shelf of the next section (1–9), in
which all the normal features are very clearly developed.
These were prepared by the late Dr. Jeffries Wyman by
exposing surfaces of plastic clay to both natural and artificial showers.

Muddy sediments, especially in lakes and rivers, are
often exposed to the air and sun during periods of drouth,
and as they gradually dry up, polygonal cracks are
formed. The sediment of the next layer will fill these
sun-cracks; and when, as often happens, it is slightly
different from the dessicated layer, they may still be
traced. The specimens from the Connecticut Valley
(22, 24) may be regarded as normal examples, except
that the sediment is somewhat arenaceous; the material
filling the cracks being, however, distinctly coarser than
that in which they were formed. In the large specimen
from Texas (23) the sun-cracks have been mainly filled
by iron oxide, and the clay subsequently washed away,
leaving the ferruginous casts of the cracks in relief. The
polished specimen from England (21) is interesting on
account of the perfect univalve shells with which the
hardened black mud or shale is crowded. The cracks in
this instance are filled with brown calcite. It is possible,
however, that this specimen is wrongly classified, and
that it really belongs to the class of concretions known as
septaria stones. (See section 13.) Sun-cracks are most

characteristic of argillaceous rocks, and, of course, prove
that in early times, as at the present day, sediments of
this class were exposed by the temporary retreat of the
water. The original surfaces of the argillaceous strata
sometimes, as in the specimen from Pennsylvania (**25**),
indicate a sluggish flowing movement of a mass of mud.

The foot-prints or trails of land animals are often, as
in the sandstones and shales of the Connecticut Valley,
associated with, and of course strongly corroborate, all
these other evidences of shore deposits. The sandstones
and shales of the Connecticut Valley have afforded an
abundance of fine examples. A single specimen, only,
is placed in the case (**61**) ; the large slab in the passage
to the Mineralogical Room and those in the Vestibule
being far more satisfactory than any that could be placed
on the shelves. It will be noticed, however, that two of
the slabs do not show the actual foot-prints, but the casts
of them instead.

Fossils.—From the foot-prints preserved in the rocks
we pass naturally to the fossil remains of both animals
and plants found entombed in the strata. Although
fossils find their highest interest in the light which they
throw upon the succession of life on the globe, they may
also be properly regarded as structural features of stratified rocks ; and any one who has seen dead shells, crabs,
fishes, etc., on the beach will readily understand how fossils get into the rocks. It is not necessary here to en-
croach upon the departments of biology or historical
geology by exhibiting the varied forms and structures of
fossils, or the fossils characteristic of the different geological formations ; but it is the province of the petrolo-

gist simply to observe and illustrate the three principal degrees in the preservation of fossils. The series of
illustrative specimens is quite complete, nearly every important phase of fossilization being represented.

1. *Original composition not completely changed.* — The
conditions existing in the stratified rocks, and especially
in the newer formations, have often proved favorable to
the preservation in whole or in part of the original composition of organic bodies. Even the most perishable
organisms are sometimes found well preserved. Thus,
mammoths (extinct elephants) have been found frozen
in the river-bluffs of northern Siberia, and so perfectly preserved that dogs and wolves ate their flesh. The bodies
of animals and of human beings have also been found
almost intact in peat-bogs after being buried hundreds of
years, the water of a bog having marked antiseptic properties. All coal is simply fossil vegetation retaining in
a large degree the original composition see the specimen (2) in the first section of this case and the same is
true of ferns (41) and other kinds of plants (44) preserved as black impressions in the rocks. The carbon of
which all such examples are chiefly composed is a part of
the original carbon of the living plants. The fossil fruits
(45) from the Tertiary formation of Brandon, Vermont,
are an admirable illustration of vegetation which has been
carbonized during its long burial in the strata, but is
otherwise unchanged. All bones and shells consist of
mineral matter which makes them hard, and animal matter which makes them tough and strong; and in very
many cases, especially in the more recent formations,
the animal matter is still partially, and the mineral

matter almost wholly, intact. Thus, in the tooth of a
mastodon (another extinct elephant) from Post-tertiary
beds (50), the cetacean bones from the Miocene strata
of Virginia (46, 48), and the teeth of gigantic extinct
sharks from the Eocene phosphate beds of South Carolina (42–43), we have still not only the original phosphate of lime (the organic form of the mineral apatite)
essentially intact, but also to some extent the original
animal matter. In the large Miocene shells (47, 49),
which may be regarded as representing a very large proportion of both fossil shells and corals, the animal matter
has entirely disappeared, but the original mineral matter
(carbonate of lime) still remains.

2. *Original composition completely changed, but form
and structure preserved.*—All kinds of fossils are commonly called petrifactions; but only those preserved in
this second way are truly petrified, *i.e.*, turned to stone.
"Petrified wood is the best illustration, and in a good
specimen, not only the external form of the wood, not
only its general structure—bark, wood, radiating silver-
grain, and concentric rings of growth—are discernible,
but even the microscopic cellular structure of the wood,
and the exquisite sculpturing of the cell-walls, are perfectly preserved, so that the kind of wood, may often be
determined by the microscope with the utmost certainty.
Yet not one particle of the organic matter of the wood
remains. It has been entirely replaced by mineral matter; usually by some form of silica. The same is true of
the shells and bones of animals."—Le Conte.

The petrified or silicified wood is well represented by
the large specimen (81) on the bottom shelf of the third

section and the smaller specimens (**1, 8**) on the top shelf of the fourth section. Especially instructive are the comparatively recent specimens from the Geyser district of the Yellowstone National Park (**7, 9**). These are still composed in part of the original woody matter ; but are well incrusted and permeated by the petrifying silica ; and illustrate the petrifying process.

We must imagine the wood as immersed in alkaline water holding silica in solution. The wood gradually becomes saturated with the solution of silica (water-logged) the water filling not only the spaces between the vegetable cells, but penetrating the cells themselves. The decomposition of the protoplasm gives rise to acids, which neutralize the alkaline solvent of the silica and cause its precipitation, each minute particle of organic matter, as it decays, being replaced by an equivalent portion of silica. Afterwards the more durable woody walls of the cells are slowly replaced by silica of slightly different texture or color and the petrifaction is complete.

A very large proportion of the shells (**5**) and corals (**6**) found in the older limestones or calcareous strata have been silicified ; and acquiring thus the hardness and durability of quartz, they are left in relief on the weathered surface of the rock. Among other petrifying substances, besides silica, are iron oxide (**2**) and iron sulphide (**3-4**).

3. *Original composition and structure both obliterated, and form alone preserved.*—This occurs most commonly with shells, although fossil trees are also often good illustrations. The general result is accomplished in several ways : (*a*) The shell after being buried in the sediment may be completely dissolved by percolating water,

leaving a *mold* of its external form. The mold or impression of an egg-shell in eolian limestone from Bermuda (**24**) is a clear illustration, although the shell has not been entirely removed in this case. The piece of highly fossiliferous rock from North Carolina (**25**) shows many thin molds of bivalve shells. The large specimen on the bottom shelf of section 3 (**82**) shows deep impressions resembling horse-tracks, which are really due to the weathering out of a fossilized marine plant; and the adjacent specimens (**83–84**) are also good examples of the molds of corals and of crinoid stems. (*b*) The mold may subsequently be filled by the infiltration of finer sediment forming a *cast* of the exterior of the shell. This phase is best illustrated by the fossil trees (**31–32**) from the Carboniferous formation, the large specimen from the celebrated section at South Joggins, Nova Scotia, which is in the Vestibule, on the left side of the main stairs, being a particularly fine example. These are trees of three different species which have been buried while standing erectly, by the rapid deposition of mud and sand ; and have then gradually decayed, leaving cylindrical holes or molds in the slowly hardening sediment, which have been subsequently filled by fine silt washed in by the water, forming these natural casts, which show so perfectly the external form of the trees, but are entirely structureless within. (*c*) The shell, before its solution. may have been filled with mud, as dead shells usually are ; and if the shell itself is then dissolved away, we have a cast of its interior enclosed in a mold of its exterior. The specimens show these very common internal casts still in the rock (**22**) and also removed from it

(21, 29-30). The section of a large ammonite shell (62, section 3) is especially interesting because only the large outer or living chamber has been filled with mud, which has hardened in it, the smaller chambers behind this all the way around the coil being partially filled with silica deposited from solution and not simply washed in as the mud was. In the smaller shell (27) of the same kind all the chambers are filled with the hardened mud, and the shell has been subsequently removed. The large univalve shell (29) is interesting because only a part of the shell has been worn or dissolved away from the cast. The crayfish incrusted with iron oxide from the water of a chalybeate spring (23), although not strictly a natural specimen, is interesting as showing another way in which molds of the exterior may be formed. The impressions of fucoids (26, 28) often observed in the strata are to be classed with the fossil trees as exterior casts.

Forms of Tufa deposits. — We have already seen that the interesting rocks known as tufas are formed when certain minerals,—especially silica, carbonate of lime, and iron oxide — are deposited from solution, not over the bottom of a sea or lake forming regular strata; but around the outlet of a mineral spring, or along the stream which it forms, or on the margin of a lake whose waters are wasting by evaporation, forming rock masses which are not plainly or regularly stratified, but present other and very characteristic structural features. These are well illustrated by the specimens of calcareous and siliceous tufas in this section. The first specimens to be noticed (23, 28, 62) are the typical examples of calca-

reous tufa, which are formed so abundantly by the incrus-
tation of vegetable forms — moss, grass, reeds, etc. ; the
subsequent decay of the organic matter leaving an exceedingly light and porous mineral network. These
specimens might have been included with the illustrations
of fossilization, the composition and structure of the
plants being entirely obliterated and the form only imperfectly preserved. The thinolite tufa from the ancient
and elevated beaches encircling the alkaline lakes of the
Great Basin (63–64) is an exceptionally interesting variety of calcareous tufa, on account of being still more distinctly pseudomorphic in character. Its seemingly crystalline structure is supposed to be due to the fact that it
was first formed by the abundant crystallization of sulphate of sodium from the waters of the lake, this salt
being afterwards replaced by carbonate of lime. A
third and quite distinct type of tufa is where the form is
original and independent of foreign objects, being determined simply by the way in which the mineral water
issues from the spring or geyser and the conditions of its
evaporation. These forms, which are extremely varied
and interesting, often imitating such organic growths as
corals, fungi, etc., and constituting one of the chief attrac-
tions of the localities in which they occur, are quite well
represented by the other specimens of calcareous tufa
from the Great Basin (61) and Vermont (49), and the
siliceous tufa from the geysers of the Yellowstone Park
(41–42, 45–47).

    **Time required for the Formation of Stratified
Rocks.**—Many attempts have been made to determine
the time required for the deposition of any given thick-

ness of stratified rocks. Of course, only roughly approximate results can be hoped for in most cases ; but these are at least sufficient to make it certain that geological time is very long. The average relative rate of growth of different kinds of sediment is, however, less open to doubt, for we have already seen that coarse sediments like gravel and sand accumulate much more rapidly than finer sediments like clay and limestone ; and we are sometimes able to compare these two classes of rocks on a very large scale.

Thus, during what is known as the Paleozoic era, a sea extended from the Blue Ridge to the Rocky Mountains. Along the eastern margin of this sea, where the Alleghany Mountains now stand, sediments—chiefly conglomerate and sandstone, with some slate and less limestone—accumulated to a thickness of nearly 40,000 feet. Toward the west, away from the old shore line, the coarse sediments gradually die out, and the formations become finer and thinner. In western Ohio and Indiana, slate and limestone predominate; while in the central part of the ancient sea, in Illinois and Missouri, the Paleozoic sediments are almost wholly limestones, and have a thickness of only 4,000 to 5,000 feet. In other words, while one foot of limestone was forming in the Mississippi Valley, eight to ten feet of coarser sediments were deposited in Pennsylvania.

The interesting specimens (**81–82**) from Silver Cay Reef, off Turk's Island, in the West Indies, showing considerable growths of coral on a large bell, olive-jar and decanter, which were recovered in 1857 from the wreck of a vessel supposed to be the British frigate Severn, lost at this place in 1793, are admirable examples of the kind of evidence upon which estimates of the time required for the formation of limestone are based. We have here in sixty-four years coral growths upwards of a foot in thickness.

But the formation of limestone strata on this reef must be much less rapid than this; for we must remember that the growth of corals on a reef is much like that of plants in a garden, isolated coral stalks and masses standing here and there and growing upwards rapidly while the intervening ground may be nearly or quite bare of living coral. When the polyps die, the brittle coral is broken up by the action of the surf and the fragments and coral sand are strewn over the general surface of the reef, which is thus gradually elevated as a whole.

The best estimates show that coral-reefs rise—*i.e.*, limestones are formed on them—at the rate of about one foot in two hundred years. But coral limestones grow much more rapidly than limestones in general. Sandstones sometimes accumulate so rapidly that trees are buried before they have time to decay and fall. This is the history of the Carboniferous tree in the Vestibule, on the left side of the main stairs. It is from the section on the South Joggins shore in Nova Scotia, where no fewer than seventy coal seams and old land surfaces, often with buried forests, have been observed alternating with marine strata. Every buried forest, like a coal-bed, represents a land surface, and proves a subsidence of the land; and in some cases repeated oscillations of the earth's crust may be proved in this way.

The mud deposited by the annual overflow of the Nile is forty feet thick near the ancient city of Memphis; and the pedestal of the statue of Rameses II., believed to have been erected B. C. 1361, is buried to a depth of nine feet four inches, indicating that 13,500 years have elapsed since the Nile began to spread its mud over the sands of the desert.

The specimen of laminated glacial clay (65) is of special interest in this connection, since it is probable that one complete layer was deposited annually. During the recession of the great ice-sheet, the melting of the ice and the resulting floods of water were intermittent, varying with the seasons. During the summer, the clay in the drift was washed down rapidly and spread over the bottom of a glacial lake, forming one of the gray layers seen in the specimen; but with the advent of the long winter the flow of water nearly ceased, the lake became covered with a thick sheet of ice; and during this period of perfect tranquillity the finest silt suspended in the water, together with any organic matter that may have been present, slowly settled, forming the brown line separating each layer of gray clay from the next. It is believed that the banded sandstones and slates (see section 1) have a similar significance, recording periodic deposition. It is not necessary, however, to suppose that the layers are always, like the rings on the section of a tree, strictly annual deposits; but they testify rather to alternating flood and drouth; or, possibly, in the case of some marine sediments, to the ebb and flow of the tide.

But the greatest difficulty in estimating the time required for the formation of any series of strata arises from the fact that we cannot usually even guess at the length of the periods when the deposition has been partially or wholly interrupted. Now and then, however, we find evidence that these periods may be very long. A layer of fossil shells in sandstone or slate (66) or on the surface of the rock (67) proves an interruption of mechanical deposition. Beds of coal, fossil forests, and other indications of land surfaces are still more conclusive. The interposition of strata (page 193) proves a prolonged interruption of deposition over the area not covered by the interposed bed. But the most important of all evidence is that afforded by unconformity (page 194); and the length of the lost interval between the two formations is measured approximately by the erosion of the older.

The old gun (**87**), the bottles (**84–85**), and other objects (**83, 86**), in this section, with oysters and other marine organisms attached, have been dredged at different points along our coast; and show very plainly that in these localities the deposition of the ordinary mechanical sediments is practically at a stand-still, for otherwise these objects would have become buried in the sand or mud before there was time for several generations of shells and other animals to grow upon them.

ORIGINAL STRUCTURES OF ERUPTIVE ROCKS.

The structures of this class are naturally divisible into those pertaining to the plutonic rocks or dikes and those pertaining to the volcanic rocks or lava-flows.

Dikes.

The term *dike* is a general name for all masses of eruptive rocks that have cooled and solidified in fissures or cavities in the earth's crust. But the name is commonly restricted to the more regular, wall-like masses, such as are represented by the first model (**1**), and by several of the specimens, and can be observed to good advantage in many of the ledges about Boston ; and the plutonic eruptions of extremely irregular outline, such as the granitic rocks usually present, as shown in the next model (**2**), are known simply as eruptive masses.

The propriety of this distinction is apparent when we consider the origin of *dike* as a geological term. It was first used

in this sense in southern Scotland, where almost any kind of a
wall or barrier is called a dike.   The dikes traverse the different
stratified formations like gigantic walls, which are often en-
countered by the coal-miners, and on the surface are frequently
left in relief by the erosion of the softer enclosing rock, so
that in the west of Scotland, especially, they are actually made
use of for enclosures.   In other cases the dike has decayed
faster than the enclosing rock, and its position is marked by a
ditch-like depression.   The narrow, straight, and perpendicular
clefts or chasms often observed on our coast are due to the re-
moval of the wall-like dikes by the action of the waves.   Dikes
are sometimes mere sheets of rock, traceable for a few yards
only; and they range in size from that up to those a hundred
feet or more in width, and traceable for scores of miles across
the country, their outcrops forming prominent ridges.   The
sides of dikes are often as parallel and straight as those of
built walls, the resemblance to human workmanship being
heightened by the numerous joints which, intersecting each
other along the face of a dike, remind us of well-fitted
masonry.

Forms of Dikes.—A dike is essentially a casting.
Melted rock is forced up from the heated interior into
a cavity or crack in the earth's crust, cools and solidifies
there, and, like a metallic casting, assumes the form of
the fissure or mold.   In other words, the form of the
dike is exactly that of the fissure into which the lava was
injected.   Now the forms of fissures depend partly upon
the nature of the force that produces them, but very
largely upon the structure—and especially the joint-
structure—of the enclosing rocks.   Nearly all rocks are
traversed by planes of division or cracks called joints,
which usually run in several directions, dividing the rock
into blocks.   And it is probable that dike-fissures are

most commonly produced, not by breaking the rocks anew, but by opening or widening the pre-existing joint-cracks. Hence dikes formed in rocks possessing a well-developed and regular joint-structure, such as the slate (3) and conglomerate (4) in the vicinity of Boston, and most kinds of sedimentary rocks, must of necessity be regular and wall-like in form; and, *vice versa*, the irregular jointing or absence of jointing often observed in gran-- ite (5) and other massive eruptive rocks gives rise to more sinuous, branching, and variable dikes.

This principle is abundantly and admirably illustrated in the vicinity of Boston, particularly in the slate quarries of Somerville, and in the large puddingstone quarry in Roxbury. The general dependence of the dikes upon the joint-structure is proved by the facts that, as may be readily observed in the quarries, the dikes, like the joints, are normally vertical or highly inclined, and that they are usually parallel with the principal systems of joints in each district. In the models (1–2) the faintly incised lines represent the joint-planes; while the stratification of the first is shown by the horizontal stripes of color. The regular and typical dikes are often branching, it is true, but in a regular and systematic manner, as shown in the model. The main dike or a branch of it may pass horizontally between the strata for a long distance, or it may approach the surface by a zigzag course, alternating in direction between the bedding-planes and joint-planes or between two sets of joint-planes.

When the dike-fissures are formed by breaking the rocks anew at the time of the eruption, it will be readily understood that if the breaking force acts slowly, so as to be influenced by all the inequalities of texture and structure, the fractures or fissures will usually be far more irregular than if the strain be developed suddenly, as during an earthquake; on the same principle that a bullet thrown by the hand will break a pane of glass more irregularly than one fired from a gun. In this way we can explain the occurrence of trap dikes of regular form in granite (**25–27**), conglomerate, and other coarse-grained rocks. The trap dikes in marble from Smithfield, R. I., (**21–23**) have the same significance; although they have been somewhat broken and faulted subsequently to their formation.

The water-worn fragment or pebble of syenite from Marblehead Neck (**24**) is divided by a tiny branching dike of trap, a small but very clear example of a dike enclosing or surrounding a relatively large mass of the bordering or wall rock.

The specimen from Mount Royal (**51**) shows two approximately parallel dikes of trap cutting obliquely across a bed of Trenton limestone, and may be regarded as illustrating the occurrence of dikes in systems after .the manner of the joint-planes. The irregularly branching forms of dikes are particularly well illustrated by the small dikes of syenite in diorite from Nahant (**82–83**) and Marblehead (**66**).

Structure of Dikes. — The rock traversed by a dike is called the *country* or *wall* rock. Fragments of this are often torn off by the igneous material and become enclosed in the latter. Such enclosed fragments partially or wholly detached are represented in certain of the dikes shown in the model (**2**). They are sometimes so numerous as to form the main part of the dike, which then, since the fragments are necessarily angular, often assumes the aspect of a breccia, in which the enclosed fragments form the pebbles and the eruptive rock the cement. The granite (**61, 81**), and syenite (**66**) eruptive through the diorite in Marblehead, Salem, and many other districts

about Boston afford magnificent examples not only of
dikes crowded with enclosed fragments, but of nearly
every kind of irregularity of both form and structure to
which dikes are subject.

The enclosed fragments in dikes sometimes throw important
light upon the relative positions and ages of the rocks traversed by the dikes. Thus the well-rounded quartzite pebble in
the coarsely crystalline trap from Somerville (**44**) proves that
although at the surface the country rock is slate, at some depth
below the surface the dike must break through beds of conglomerate from which the melted trap has picked this pebble on its
way up through the fissure, and hence that the conglomerate
probably underlies the Somerville slate. Other instances of the
same kind have been observed at Nantasket.

The enclosed fragments afford the only important exception to the rule that dikes are homogeneous in composition ; i.e., in the same dike we can usually find — from
end to end, from side to side, and probably from top to
bottom — no essential difference in composition. This
homogeneity is well illustrated by several of the specimens. But there is often a marked contrast in *texture*
between different parts of a dike and especially between
the sides and central portion. The liquid rock loses heat
most rapidly where it is in contact with the cold walls of
the fissure, and solidifies before it has time to crystallize,
remaining compact and sometimes even glassy ; while in
the middle of the dike, unless it is very narrow, it cools
so slowly as to develop a distinctly crystalline texture.
There is no abrupt change in texture, but a gradual passage from the compact border to the coarsely crystalline
or porphyritic middle portion. It is obvious that a sim-

ilar gradation in texture must exist between the top and bottom of a dike. It is difficult to observe this gradation between the wall and center in small dikes, because they are essentially compact throughout, but it is slightly indicated in one of the little dikes from Marblehead (25) ; and the two hand specimens of trap (52–53) represent the wall and center portions respectively of a dike forty feet wide on Marblehead Neck.

Enclosed fragments of the wall rock have sometimes exerted a similar chilling influence upon the contiguous igneous rock ; so that they are immediately enveloped by a layer which is much more compact than the general mass. This is very clearly shown by the quartzite pebble already referred to (44) in the coarse Somerville trap. It is usually observed, also, that while the main part of a large dike is distinctly or coarsely crystalline, the small, branch dikes running off from it are of a very compact texture. This is well illustrated by the two specimens from a large dike in Somerville. The black, coarse-grained specimen (46) represents the main part of the dike, and the small, fine-grained or compact dike in slate (49) is one of the narrow branches extending out into the slate. Dikes, especially near the original surface; or where they have been formed under little pressure, are sometimes vesicular (28), after the manner of ordinary lavas ; or the steam-holes may be filled with secondary minerals, making the rock amygdaloidal (29).

Somewhat related to the gradation in texture is the flow-structure parallel with the walls occasionally observed in dikes. This structure, which sometimes takes the form of a distinct banding or striping, is well developed in some of the dark, crystalline diabase of Nahant and the outer islands of Boston Harbor (65). It clearly indicates that the igneous rock continues to flow after it has begun to cool and crystallize, and that the fissure is gradually narrowed by the formation on either wall of successive layers of solid trap.

One of the most interesting and important structures of plutonic rocks is that illustrated by the two large and seemingly stratified masses of syenite from Marblehead (**62–63**). The syenite, which is the coarsely crystalline, feldspathic rock (light-colored), is eruptive here through the finely crystalline diorite (dark-colored). The diorite is itself an older eruptive rock in which, as the result of enormous pressure, possibly during its original solidification, an imperfect cleavage or foliation had been developed. This structure, which resembles stratification, causes the diorite to split easily in parallel vertical planes ; and the liquid syenite instead of forming and filling one wide fissure in the fissile diorite, has, under great pressure, been injected in thin layers along all these planes of weakness, thus giving rise to a very perfect interlamination of the two rocks ; although, as the specimens show, the syenite sometimes breaks across the diorite, forming ordinary dikes. When the diorite is less fissile, the relations of the two rocks are represented by the smaller specimen (**66**) and by the other examples of enclosed fragments from Marblehead (**61, 81**), except that the injected rock is mainly granite instead of syenite.

Truly stratified rocks are also sometimes injected along the bedding-planes in this intimate way by igneous material; and there can be no doubt but that many of the so-called gneisses or crystalline stratified rocks are partly if not wholly of igneous origin.

Contact Phenomena. — Under this head are grouped the interesting and important phenomena observable along the contact between the dike and wall-rock. These throw light upon the conditions of formation of dikes, and are often depended upon to show whether a rock mass is a dike or not. We may notice here : —

1. *The detailed form of the contact.* It may be straight and simple as in the first model (**1**) and several of the specimens (**25–27, 51**) or exceeding irregular, the dike penetrating the wall, and enclosing fragments of it, as in the second model (**2**), which represents a typically igneous contact, and as in most of the specimens already noticed, and particularly in the contact of trap and marble from Lewiston, Me., (**43**)

2. *The alteration of the wall-rock by heat.* This may consist in : (*a*) *Coloration*, shales (**84**) and sandstone (**85**) being reddened in the same way as when clay is burnt for bricks, or whitened by the oxidation of the carbon in carbonaceous clays. (*b*) *Baking and induration*, sandstone being converted into quartzite (**85**) with the hardness and color of jasper ; clay, slate, etc., being not only baked to a flinty hardness (**86–87**), but actually vitrified, as in porcelainite ; and bituminous coal being converted into natural coke (**88**) or anthracite. (*c*) *Crystallization*, chalk (**89**) and other forms of limestone being changed to marble, and crystals of pyrite (**90**), calcite, quartz, etc., being developed in slate, sandstone and other rocks, through the direct action of the volcanic heat or of the thermal waters accompanying the eruption. The massive garnet and other minerals in the calcareous

slates of Nahant (**91**), and Newbury (**92**), represent a
large number of silicate species which owe their origin to
contact metamorphism and crystallization.

3. *The alteration of the dike rock.* (*a*) By the more rapid
cooling near the walls, as already explained (**41–42**),
rendering it more compact in texture. (*b*) By the access
of thermal or meteoric waters, which may decompose the
eruptive rock, as in the case of the small branch dikes in
the Somerville slate (**45, 47–50**).

The alteration of the wall-rock may extend only a few
inches or many yards from the dike, gradually diminishing with the distance; and the cases are very numerous
where there is no perceptible alteration (**51**, etc.) ; and,
again, as just explained, the alteration is usually mutual,
the dike rock being altered in texture, color, and composition.

Intrusive Beds. — We commonly think of dikes as
cutting across the strata, but they often lie in planes
parallel with them; and the same dike may run across the
beds in some parts of its course and between them in
others, as shown in the model (**1**) ; or the conformable
dike may be simply a lateral branch of a main vertical
dike, as may be seen in the same model; or, finally, a
dike cutting across the strata may end abruptly at a particular plane of stratification and spread out between the
strata. All dikes or portions of dikes lying conformably
between the strata are called *intrusive beds* or *sheets.*

Intrusive beds are exposed at many points in the vicinity of
Boston. The great mass of diabase forming the principal part
of Nahant is an intrusive bed in the slate; but the clearest and
most typical examples occur in the slate formations of the outer

Islands of Boston Harbor.   The picture (64) shows a natural
section on the south side of Middle Brewster Island, the lighter
color representing the slate and the darker the interbedded
sheets of diabase.

When a dike, failing to reach the surface, spreads out
horizontally between the strata, forming a thick dome-
shaped intrusive bed, *i.e.*, an intrusive bed of great thickness in proportion to its horizontal extent, it is called
a *laccolite*.   The model (2) gives the general idea of a
laccolite, as seen in section and in relief ; and also affords
a comparison between a laccolite and a volcano.   In
the one case a large mound of eruptive material accumulates between the strata, the overlying beds being
lifted into a dome ; while in the other case the fissure or
vent reaches the surface, and the mound of lava is built
up on top of the ground.   Laccolites are sometimes of
immense volume, containing several cubic miles of igneous rock.   The laccolites first described, and one of the
largest and most typical groups yet discovered, are those
forming the Henry Mountains, in Utah.   These are well
represented by the two relief maps in the middle window-
space of this room.   The first map shows the laccolites as
they probably appeared when first formed.   The overlying strata are lifted into smoothly rounded domes which
completely conceal the igneous masses, and the uniform
surface of the entire map, the complete absence of evidences of erosion, indicates that this region had not been
elevated above the sea since the deposition of these Cretaceous beds.   The companion map, on the other hand,
shows the present appearance of the same area, after extensive erosion has removed a large part of the newer

.strata, developing the drainage systems, and exposing the summits of the laccolites.

The sedimentary beds (Cretaceous) can, however, still be seen arching up over the slopes of the igneous cores on every side. It is, in fact, this circumstance that most clearly distinguishes a laccolite from a true volcano; for even if the latter were submerged and covered by sedimentary deposits, they would rest horizontally and unconformably against its slopes, and not arch regularly and conformably over it. Laccolites are more numerous than they were formerly supposed to be, many eruptive masses which were once classed as volcanoes being now regarded as laccolites.

Ages of Dikes. — The ages of dikes may be estimated in several ways. They are necessarily newer than any rock formations which they intersect or of which they enclose fragments. But any sedimentary strata crossing the top of a dike must usually be regarded as newer than the dike, especially if they contain water-worn fragments of the dike rock.

The relative ages of different dikes are most easily and satisfactorily determined by their intersections, on the principle that when two dikes cross each other, the intersecting must be newer than the intersected dike. This principle is illustrated by several dikes in the model (**1**), some of the intersections showing in both the surface and sectional views ; and it is apparent that it must sometimes be possible, in this way, to prove several distinct periods of eruption in the same limited district.

The textures of dikes also often afford reliable indications of their ages ; for, as we have already seen, the upper part of a

dike, cooling rapidly and under little pressure, must be less
dense and crystalline than the deep-seated portion, which cools
slowly and under great pressure. Now, the lower, coarsely
crystalline part of a dike can usually be exposed on the surface
only as the result of enormous erosion; and erosion is a slow
process, requiring vast periods of time. Hence, when we see a
coarse-grained dike outcropping on the surface, we are justified
in regarding it as very old, for all the fine-grained upper part
has been gradually worn away by the action of the rain, frost,
etc. Other things being equal, coarse-grained must be older
than fine-grained dikes; and the texture of a dike is at once a
measure of its age and of the amount of erosion which the
region has suffered since it was formed.

Eruptive Masses. — In striking contrast with the more or
less wall-like dikes are the highly irregular, and even ragged,
outlines of the eruptive masses; and it is worth while to notice
the probable cause of this contrast. The true dikes are formed,
for the most part, of comparatively fine-grained rocks — the
typical "traps"; while the eruptive masses consist chiefly of the
coarse-grained or granitic varieties. Now we have just seen
that the coarse-grained rocks have been formed at great depths
in the earth's crust, while the fine-grained are comparatively
superficial. But we have good reason for believing that the
joint-structure, upon which the forms of dikes so largely
depend, is not well developed at great depths, where the rocks
are toughened, if not softened, by the high temperature. In
other words, trap dikes are formed in the jointed formations,
which break regularly; while the granitic masses are formed
where the absence of joint-structure and a high temperature
combine to cause extremely irregular rifts and cavities when the
crust is broken. And we may suppose that plutonic masses
which are coarsely crystalline and extremely irregular in form
at great depths in the earth often pass gradually upward into
ordinary fine-grained, wall-like dikes.

Volcanoes.

Volcanic eruptions, or the actual emission of lava at the earth's surface, are of two distinct types : (1) fissure eruptions, where the lava issues from a fissure or series of fissures, often in enormous volume, and forms broadly extended sheets and beds of lava, sometimes of great thickness, as in the northwestern part of the United States ; (2) crater eruptions, where the lava issues from a more circumscribed vent or crater and builds up an ordinary volcanic cone. The fissure eruptions, although of vast importance in the structure of the earth, are evidently not adapted to museum illustration ; but the remaining specimens and models under the general head of the eruptive rocks illustrate some of the varied structural features of crater eruptions.

Volcanic Cones. — The crater eruptions, as explained on page 43, may be distinguished as (1) *quiet*, when the lava issues mainly in a liquid form ; or (2) *explosive*, when it is largely blown out in the form of dust and fragments. The general forms of volcanic cones evidently depend largely upon the relative proportions of liquid and solid lava, the former making gentle and the latter steep slopes. This contrast may be observed in the large relief map of Mt. Vesuvius, in the Vestibule. The lower part of the mountain, built up largely of successive flows of liquid lava, is much less steep than the upper cone, in which the volcanic ashes and cinders largely predominate. The extreme examples are seen in some of the smaller cinder cones, such as Monte Nuovo, near

Naples, rising at angles of more than thirty degrees, and
the great volcanoes of the Sandwich Islands, which,
although in some cases more than 14,000 feet high, are so
broad that the angle of slope is almost inappreciable.
The relief map of Mt. Vesuvius, already referred to,
shows, in Monte Somma, part of the wall of an older and
larger crater, encircling the modern cone, and thus illustrates the concentric craters which are frequently developed when minor eruptions follow greater ones. The
ideal section of a volcanic cone (1) shows outward sloping layers, representing the successive flows of lava in the
history of a volcano ; the lateral eruptions, with the dikes
and monticules which they form ; monticules which have
been buried by subsequent eruptions from the main
crater ; the underlying sedimentary rocks on which the
volcano stands ; and the neck or pipe extending down
through these to the original source of the lava.

In like manner, the sections on the next model (2)
illustrate the different phases of extinct and fossil volcanoes, showing cones partially worn away ; partially and
completely buried by sedimentary deposits ; cones that
have been tilted by subsequent disturbance and folding of
the strata ; and the various appearances of the volcanic
pipes or necks, where the cone itself, the surface accumulation of solid and fragmental lava, has been completely
worn away. Some of the characteristic appearances of
an extinct volcano are also well illustrated by the smaller
models (3-4) and by the photographs (5) ; but no explanations are required beyond those given on the labels.
The large relief map of the island of Oahu (section 7)
shows, in the main mountain ridges, remnants only of

two gigantic volcanoes, the higher central portions of
which have been completely swept away by subaerial and
marine erosion. The small cones, rising from the lower
land, and distinguished by having the crater-rims painted
red, were formed by very much later and relatively insignificant eruptions, largely of the explosive type. In the
same connection, also, we may profitably refer to the
large models in the upper part of sections 12 to 16.

Minor Structures of Lava. — The specimens in
this section (6) illustrate, necessarily, only some of the
minor features of recent lava. These are interesting especially for the very clear evidence which they afford of
the former liquidity of the lava. This evidence is particularly striking in the stalactitic and stalagmitic forms
shown in the first group **(21–29)**.

This fine series is from the crater of Kilauea on the Island of
Hawaii. The lava of the Hawaiian volcanoes is exceptionally
liquid, and when it drips from overhanging surfaces the long,
stringy drops build up mounds or slender columns. During
their exposure to the air before and while falling the liquidity
of the drops is sufficiently reduced by cooling so that they remain more or less distinct in the columns, being just viscous
enough to stick together and yet stiff enough to retain their
individual forms, presenting distinctly worm-like aspects. All
these columns built up of successive drops of lava are, of
course, essentially stalagmitic in character. But the slender
cylindrical specimens, not so evidently composite in structure,
are true stalactites, having resulted from the drawing out by the
action of gravitation, of a single large drop of lava. The steam
dissolved in the liquid lava is liberated during its solidification
and renders all these forms more or less cellular or hollow, some
of them being light enough to float on water. These bubbles of
steam or steam-holes are formed, not alone in isolated drops of

lava, but very generally in the superficial portions of a lava stream, where the diminished pressure permits the expansion of the steam. Where the consistency of the lava is most favorable, they may attain a considerable size, forming prominent tumefactions or swellings on the surface of the flow.

The large specimen from the Azore Islands (47) is undoubtedly an irregular stalagmitic mass. The bright red color is due to the peroxidation of the iron oxide in the lava since its exposure to the air. This change in color of the darker lavas is very commonly observed, especially in warm countries. The other specimens on this shelf show the more ordinary scoriaceous forms of basic lavas, or those that have been highly liquid ; and in some of these the flowing of the mass is very obvious. The broadly flowing, ropy and wave-like forms are illustrated by the large specimen from Mt. Vesuvius (83). This specimen is from the surface of the great flow of 1872.

In the older lavas, the steam-holes and other cavities in the lava have usually been filled by secondary minerals, developing the amygdaloidal structure, as seen in the ancient basaltic lava or melaphyr from the vicinity of Boston (85).

Volcanic Bombs. — The specimens on the fourth shelf (61–66) are examples from different volcanic regions of volcanic bombs, that is, small masses of lava which have been thrown with explosive violence from the crater into the air while in a liquid or pasty condition. During their rotation in the air they naturally assume a rounded form and harden sufficiently by cooling to retain this form after striking the ground. The whirling of the plastic mass is especially obvious in the large bomb from Auvergne (66).

Most of the specimens show indentations and cracks formed in part by cooling but chiefly by the concussion on striking the ground; and it is clearly impossible to draw any sharp line of distinction between typical bombs and ordinary clinkers or fragments of lava which reach the ground in a sufficiently pasty condition to stick or even flow together. The specimens, it will be noticed, are all basaltic, with a single exception (63). This represents the obsidian, or acid and difficultly fusible lava, of Vulcano. Its imperfect fluidity at the time of eruption is seen in the subangular form of the mass and in the cracks, ("bread-crust" structure), which are undoubtedly due to cooling.

Flow-structure. — One of the most characteristic and important structural features of igneous rocks is the flow-structure, or the fluidal lines due to the flowing of the material, and especially to its continued flowing after it began to solidify. This structure is seen alike in surface flows (82-84) and in dikes and plutonic masses (81); and it frequently simulates stratification very closely. The block of obsidian from the Lipari Islands (84) is a splendid example of flow-structure in a recent lava; while the specimen from Milton, Mass., (82) shows that it may be equally perfect and well-preserved in the ancient, devitrified obsidian known as felsite. The slab of trachyte (81) showing broad fluidal lines or bands is from Terry's Peak, an ancient laccolite in the Black Hills. And the possibility of flow-structure in ordinary dikes and plutonic masses is apparent from the specimen (64, section 5) of diabase from the vicinity of Boston.

Contemporaneous Beds. — When the lava emitted by a crater is sufficiently liquid, it spreads out horizontally, forming a volcanic sheet or bed. If such an eruption is submarine, or the lava-flow is subsequently covered by the sea, sedimentary deposits are formed over it ; and beds of lava which thus come to lie conformably between sedimentary strata are known as *contemporaneous sheets* or *beds* ; because they belong, in order of time, in the position in which we find them, being, like any member of a stratified series, newer than the underlying and older than the overlying strata. Contemporaneous lava-flows are sometimes repeated again and again in the same district, and thus important formations are built up of alternating igneous and aqueous deposits. Very typical examples of this sort occur in different parts of the Boston Basin, as at Nantasket and Brighton, beds of melaphyr and porphyrite alternating with beds of conglomerate and sandstone. Dikes, as we have seen (page 220), are often conformable with the strata and, evidently, the student who would read correctly the record of igneous activity in the past must be able to distinguish intrusive from contemporaneous beds.

The principal points to be considered in making this distinction are : (1) The intrusive bed is essentially a dike, dense and more or less crystalline in texture, altering, and often enclosing fragments of, both the underlying and overlying strata, and frequently jogging across or penetrating the sediments. (2) The contemporaneous bed, on the other hand, being a lava-flow, is much less dense and crystalline, but is usually distinctly scoriaceous or amygdaloidal; especially at the borders, and the underlying strata alone show heat action, or occur as enclosures in the lava; for the overlying strata are newer than the lava, and often consist largely, at the base, of water-worn fragments of the lava.

ORIGINAL STRUCTURES OF VEIN ROCKS.

Many things called veins are improperly so called, such as dikes of granite and trap, and beds of coal and iron ore. The smaller, more irregular, branching dikes, especially, are very commonly called veins, and to distinguish the true veins from these eruptive masses, they are designated as *mineral veins* or *lodes*, although the term *lode* is usually restricted to the metalliferous veins.

Origin of Veins. — Various theories of the formation of veins have been proposed, but most of these are of historic interest merely, for geologists are now well agreed that nearly all true veins have been formed, as briefly explained on page 48, by the deposition of minerals from solution in fissures or cavities in the earth's crust. In many cases, especially where the veins are of limited extent, it seems probable that a part or all of the mineral matter was derived from the immediately enclosing rocks, being dissolved out by percolating water; and these are known as segregation or lateral secretion veins. But it is quite certain that as a general rule the mineral solutions have come chiefly from below, the deep-seated thermal waters welling up through any channel opened to them, and gradually depositing the dissolved minerals on the walls of the fissure as the temperature and pressure are diminished. This case, however, differs from the first only in deriving the vein-forming minerals from more remote and deeper portions of the enclosing rocks; and thus we see that vein-formation, whether on a large or a small scale, is always essentially a process of segregation. We know that every volcano and every lava-flow must be connected below the surface with a dike; and it is almost equally certain that the waters of mineral springs forming tufaceous mineral deposits on the surface, as in the geyser districts, also deposit a portion of the dissolved minerals on the walls of the subterranean channels, which are thus being gradually filled up and converted into mineral veins, which will be exposed on the surface when erosion has removed the tufaceous overflow. This connection of vein-formation with the superficial deposits of existing springs has been clearly proved in several important instances in Nevada and California.

Veins occur chiefly in old, metamorphic, and highly disturbed formations, where there is abundant evidence of the former existence of profound fissures, and in regions similar to those in which thermal springs occur to-day.

External Characteristics of Veins. — The typical vein, in its simplest form, is well represented by several specimens in the upper part of section 7[1], straight and regular cracks in slate and other rocks having been filled by the infiltration of calcite and quartz. In other words, a typical vein may be described as a fissure of indefinite length and depth filled with mineral substances deposited from solution. Externally it is very similar to the typical dike, for the fissures are made in the same way for both. Veins are normally highly inclined to the horizon; they

---

[1]Sections 7 and 8 are to be regarded as one, with one series of numbers; the odd decades on the left and the even decades on the right.

exhibit in nearly every respect the same general relations
as dikes to the structure of the enclosing or country rock ;
and the ages of veins are determined in the same way as
the ages of dikes.

Other specimens in this section show less **regular** veinlets of quartz and calcite in slate, sandstone, and marble,
illustrating on a small scale the branching and other normal and common irregularities of veins.

Veins are the chief repositories of ores; and the extensive
mining operations to which they have been subjected in all parts
of the world, have made our knowledge of their forms below
the surface very full and accurate. It has been learned in this
way that very often the corresponding portions of the walls of
a vein do not coincide in position, but one side is higher or
lower than the other, showing that the walls slipped over each
other when the fissure was formed or subsequently; and this
faulting or displacement of the walls appears to be much more
common with veins than with dikes, perhaps because the fissures remained open much longer. This differential movement
of the walls is the principal cause of the almost constant
changes in the width of veins. For, since the walls are never
true planes and are often highly irregular, any slipping of one
past the other must bring them nearer together at some points
than at others. This occurs chiefly with large veins, and hence
is not easily shown in a collection. As a rule, the enormous
friction accompanying the faulting either crushes the wall-rock,
or polishes and striates it, producing the highly characteristic
surfaces known as *slickensides*; a feature which is commonly
observed with joint-structure and will be fully-illustrated in
that connection, only two small slickensided veins of quartz
(**28-29**) being shown here. When the wall is finely pulverized
in this way, or is partially decomposed before or after the filling of the fissure, a thin layer of soft, argillaceous material is
formed, separating the vein proper from the wall-rock. The

miners call this the *selvage*; and it is a very characteristic feature of the large fissure veins, but is rarely observed in the smaller examples, such as must be used for illustrations.

Fragments of the wall-rock are frequently enclosed in veins, as may be seen in several of the specimens already referred to and especially in the irregular vein of calcite in trap (13) and the piece of lead and zinc ore from Kansas (14), and the veins sometimes branch or divide in such a way as to enclose a large mass of the wall, which is known as a "horse" (4). A similar result is accomplished when a fissure is reopened after being filled, if the new fissure does not coincide exactly with the old. This also is a feature to be observed chiefly with large veins; but something of it can be seen on the left side of the beautiful section of a rhodochrosite and quartz vein from Montana (22). It has been proved that veins have thus been reopened and filled several times in succession; and in this way fragments of the older vein material become enclosed in the newer.

Although usually determined in direction by the joint-structure of the country-rock, veins are often parallel with the bedding, especially in highly inclined, schistose formations. Such interbedded veins are commonly distinctly lenticular in form, occupying rifts in the strata which thin out in all directions and are often very limited in extent. The little vein of quartz in mica schist (15) is in nearly every respect a typical example. Among the illustrations of dikes (page 218) we have seen that the eruptive material has in some cases penetrated in thin sheets along all the bedding-planes of thin-bedded, shaly or foliated rocks, giving rise to a minute interlamination of rocks radically distinct in origin. A precisely similar relation is often observed between veins and the enclosing rocks. The specimen

from Somerville (**45**) shows such sheeted veinlets in a trap rock
which has been foliated by the development of slickensides in
it; while the mica schist (**46**) shows the same thing for a foliated sedimentary rock; and the streaks of slate in the section
of the auriferous quartz vein from Nova Scotia (**44**) require a
similar explanation.

Whether conforming with the joint-structure or bedding,
veins are commonly arranged in systems by their parallelism, those of different systems or directions usually differing in age and sometimes in composition, and the
older veins being generally faulted or displaced when
intersected by the newer. The small veins of calcite in
slate from Somerville (**11**), and of fibrous gypsum in
earthy gypsum (**16**), are examples of parallel veins; and
the polished specimen from Vermont (**24**) is a very neat
illustration, on a small scale, of two distinct systems of
veins.

In this instance, however, the systems are of the same age
and composition, the two series of cracks in the black slate
having been filled simultaneously with finely fibrous calcite.
The fissures in this specimen are intersecting, but the veins are
not strictly so.

Internal Characteristics of Veins.— Internally,
veins and dikes are strongly contrasted; and it is upon
the internal features, chiefly, as previously explained,
that we must depend for their distinction. In metalliferous veins the minerals containing the metal sought for
(the galenite, sphalerite, etc.) are the *ore*; while the
non-metalliferous minerals (the quartz, feldspar, calcite,
etc.) are called the *gangue* or vein-stone proper.

This distinction is illustrated by several of the specimens. One of these (21) is a polished section of an
English lead vein, in which the purple fluorite and yellow
barite forming the main part of the vein are the gangue
and the ore includes only the dark masses of galenite
(lead ore) in the middle of the vein. The section of a
vein from Pennsylvania (23) shows both lead ore and
zinc ore with a quartz gangue.

Although the combinations of minerals in veins are almost
endless, yet certain associations of ores with each other and
with different gangue minerals are tolerably constant, and constitute an important subject for the student of metallurgy and
mining.

When a vein is composed of a single mineral, as quartz or
calcite, it may rival a dike in its homogeneity. Most important
veins, however, are composed of several or a large number of
minerals, which may be sometimes more or less uniformly mixed
with each other, but are usually distributed in the fissure in a
very irregular manner. The great granite veins which are
worked for mica, feldspar, and quartz, are good illustrations, on
a large scale, of the structure of veins in which several minerals
have been deposited contemporaneously. The individual minerals are found to a considerable extent, in large, irregular masses,
with no order observable in their arrangement.

When a mineral is deposited from solution, it crystallizes by
preference on a surface of similar composition, thus quartz on
quartz, feldspar on feldspar, and so on; and it seems probable
that this selective action of the wall-rock may be a principal
cause of the irregular distribution of minerals in veins. It has
often been observed in metalliferous veins that the richness
varies with the nature of the adjacent country rock. This dependence of the contents of a fissure upon the wall-rock may be
due in part to the selective deposition of the minerals, and in
part to their derivation from the contiguous portions of the

country or wall-rock, as in the so-called segregated veins. Temperature and pressure exert an important influence upon chemical precipitation, and it is, therefore, probable that the composition of many veins varies with the depth.

The most important structure observed in veins is the banding or appearance of stratification which is produced when the mineral matter is deposited in more or less regular layers over the walls of the fissure. This banding may sometimes be observed even when the vein is composed entirely of one mineral, the vein then presenting a double appearance. The little vein of chalcedony from the Bad Lands **(25)** is an example. The specimen shows the entire thickness of the vein, which was formed by the deposition of silica on each side of a crack in the Tertiary marl, the deposition continuing until the two layers thus formed met in the middle in some parts of the fissure; but the deposition of the silica has been followed in the wider parts of the fissure by calcite, giving three distinct bands in all, the calcite forming one double band, and showing what may be regarded as the normal irregularity. Frequently, perhaps usually, the minerals of composite veins, as in this instance, are deposited in succession, instead of contemporaneously, and the number of bands is increased. The first mineral deposited in the fissure forms a layer covering each wall, and is in turn covered by layers of the second mineral, and that by the third, and so on, until the fissure is filled or the solution exhausted. Thus, in the vein of coarse granite from Fitchburg **(30)** we observe between the walls of fine grained granite, on either side, a layer of white feldspar with a little mica and tourmaline, while the middle of the fissure is filled with a solid (double) layer of quartz. And in the specimen from a vein of lead and zinc ore, from Pennsylvania (23), we have a complete section showing on each side, first a layer of zinc ore (sphalerite), second a layer of quartz containing some crystalline masses of lead ore (galenite), and then down the middle of the vein a solid layer of finely granular galenite. Still more striking is the polished section of an auriferous vein from Montana (22). The first mineral deposited here, evidently, was quartz containing some pyrite, the layer of quartz on one side, as already observed, having been subsequently broken up. Then came in succession more pyrite, the broad bands of rhodochrosite, and narrower bands of quartz. The latter completed the vein at most points, but in the wider parts cavities were still left in which more rhodochrosite was afterwards deposited. This is a very instructive specimen, showing well, among other things, how the normal regularity of the banding is influenced or disturbed by the tendency to crystallization and segregation.

The regular alternation of minerals in a vein is, however, most beautifully illustrated by the polished section from England (21). We can count in this, on each side of the middle, no fewer than six layers of fluorite (purple and clear) alternating with five layers of barite (creamy white and buff) ; while a single broken band of galenite marks the central line or axis of the vein.

The barite bands are somewhat interrupted, especially on the right side ; and in this respect and also in the rounded or mammillary outlines of the detached masses, they afford another excellent illustration of the way in which the tendency of the solution

to form continuous layers of each mineral is opposed by the
tendency inherent in the mineral itself to grow independently
of the surface. This principle has a very wide application in
geology; for molecular and mechanical forces are often at variance and each modifies the action of the other.

Slender or prismatic crystals occurring in veins are
usually perpendicular to the walls of the vein ; and a
layer made up in this way of transverse crystals is called
a *comb*. The large specimen of quartz (**41**) is an excellent illustration of a comb, being part of one side of a
quartz vein ; and the next specimen (**42**) representing a
more complex and somewhat broken vein of quartz, illustrates various phases of *comb-structure*, including both
single and double combs. In the vein of calcite from
Isle Royal (**12**) the crystals are distinctly prismatic, and
form two combs, separated by a thin layer of iron oxide.
Since the crystals grow from the wall of the vein toward
the middle, the inner surface of each comb must show the
free, growing ends of the crystals, as seen in the specimen
just referred to, and it is easy to see that the crystals of
one comb must often project into or even through the
layer or comb formed next after it, successive combs
being locked together in this way. In the second granite
vein from Fitchburg (**43**) something of this can be seen,
slender crystals of tourmaline starting from the walls and
penetrating the subsequently, formed layers of feldspar
and quartz. Still better is the granite vein from Chester-
field on the bottom shelf (**82**) in which slender crystals
of red and green tourmaline pass through the well-defined
layers of lamellar albite .(clevelandite) into the layer of
smoky quartz forming the middle of the vein.

This tendency of crystals of prismatic habit to grow perpendicularly to the walls of the vein is seen also in the case of the extremely slender or fibrous minerals, such as asbestus, satin spar (27), chrysolite, etc., the fibers being arranged cross-wise in the vein, and the length of the fibers measuring the width of the vein. One of the specimens referred to s a distinctly banded vein, showing two layers or combs.

The various specimens show clearly how the banding of veins may be distinguished from the stratification of the sedimentary rocks.   The main points to be observed are the mineral composition, the relations to the enclosing rocks, the repetition of the bands in reverse order on opposite sides of the middle of the vein, and the comb-structure.   The banded structure of veins is exactly reproduced in miniature in the banding of agates, geodes, and the amygdules formed in the steam-holes of old lavas.

The unfilled cavities which frequently remain along the middle of a vein (42) are called *vugs* or *pockets*. As in the case of geodes, they are commonly lined with crystals, and when the latter are minute, the pockets are called *druses*.   In metalliferous veins, the ore is often much more abundant in some parts than in others, and these *ore-bodies* or *pay-streaks*, especially when somewhat definite in outline, are known in their different forms and in different localities, as *courses*, *slants*, *shoots*, *chimneys*, and *bonanzas* of ore.   The intersections and junctions of veins are often among the richest parts, as if the meeting of dissimilar solutions had determined the precipitation of the ore.   But these features, evidently, do not readily admit of museum illustration.

Metalliferous veins, especially, are usually deeply decomposed along the outcrop by the action of atmospheric

agencies.   The ore is oxidized, and to a large extent removed by solution, leaving the quartz and other gangue
minerals in a porous state, stained by oxides of iron,
copper, and other metals, forming the *gossan* or *blossom-
rock* of the vein.

Peculiar Types of Veins and Ore-Deposits. — In
calcareous or limestone formations, especially, the joint-
cracks and bedding-cracks are often widened through
the solution of the rock by infiltrating water, and thus
become the channels of a more or less extensive subterranean drainage, by which they are more rapidly enlarged
to a system of galleries and chambers, and, in some
cases, large limestone caverns.   The water dripping into
the cavern from the overlying limestone is highly charged
with carbonate of lime, which is largely deposited on the
ceiling and floor of the cavern, forming stalactitic and
stalagmitic deposits.   These are masses of mineral matter
deposited from solution in cavities in the earth's crust,
and are essentially vein-formations.

It appears best, therefore, to class stalactites (**31–36, 71–73**)
and stalagmites (**51–53**) as special structural features of veins.
But the specimens require no particular explanation, beyond
what is given on the labels.   Portions of caverns deserted by
the flowing streams by which they were excavated, are often
filled up in this way, being converted into irregular veins of
calcite.   But calcite is not the only mineral found in these
cavern deposits; for barite and fluorite and various lead and
zinc ores, especially the sulphides and carbonates of these
metals, have also been leached out of the surrounding limestone and concentrated in the caverns.   The celebrated lead
mines of the Mississippi Valley are of this character.   The
forms of these cavern-deposits vary almost indefinitely, and are

often highly irregular.   The principal types are known as *gash-veins*, *flats* and *sheets*, *chambers*, and *pockets*.

Where joints and other cracks have opened slightly in different directions and become filled with infiltrated ores, we have what the German miners call a *stockwerk*, — an irregular network of small and interlacing veins.   The small veinlets of calcite in slate from Vermont (**24, 26**) may be regarded as forming miniature stockworks.

An *impregnation* is an irregular segregation of metalliferous minerals in the body of some eruptive or massive rock, usually along a joint-crack or a fault-plane.   Its outlines are not sharply defined, but it shades off gradually into the enclosing rock.   We must consider that the crack or fissure affords a channel for the passage of mineral waters and that the dissolved minerals, instead of simply forming a narrow vein in the crack, impregnate the solid rock on either side, the substance of the rock being often to a large extent or wholly dissolved and replaced during the process.

This process of substitution or metasomatosis is now regarded as of vast importance, many of the principal and most valuable ore deposits requiring this explanation.   The specimen of slate from Quincy (**17**), which has been bleached and perhaps otherwise altered along two intersecting joint cracks, is not properly an impregnation, since it is probable there has been simply a leaching out of material without any important addition; but it serves to illustrate the general idea of the alteration of the rock along dynamic cracks or planes of weakness.

Fahlbands are similar ill-defined deposits or segregations along the bedding-planes of stratified rocks.   The

pyritiferous gneiss from Rowe (**18**) is a general illustration. The pyrite is quite certainly not an original constituent of the gneiss, and yet it does not form distinct
veins or veinlets ; but it is disseminated through the rock
parallel with the bedding—a stratified impregnation, or
fahlband. An impregnation, vein, or other form of ore-
body occurring along the contact between two dissimilar
rocks is called a *contact deposit*. These are usually found
between formations of different geological ages, and
especially between eruptive and sedimentary rocks.

SUBSEQUENT STRUCTURES PRODUCED BY SUB-TERRANEAN AGENCIES.

The subterranean forces concerned in the formation of
rocks are chiefly various manifestations of that enormous
tangential pressure developed in the earth's crust, partly
by the cooling and shrinking of its interior, but largely,
it is probable, by the diminution of the velocity of the
earth's rotation by tidal friction, and the consequent
diminution of the oblateness of its form. It is well
known that the centrifugal force arising from the earth's
rotation is sufficient to change the otherwise spherical
form of the earth to an oblate spheroid, with a difference
of twenty-six miles between the equatorial and polar diameters. It is also well known that while the earth turns
from west to east on its axis, the tidal wave moves around
the globe from east to west, thus acting like a powerful
friction-brake to stop the earth's rotation. Our day is
consequently lengthening, and the earth's form as gradually approaching the perfect sphere. This means a very
decided shortening and consequent crumpling of the

equatorial circumference, and is equivalent to a marked shrinkage of the earth's interior, so far as the equatorial regions are concerned.

The most important and direct result of the horizontal thrust, whether due to cooling or tidal friction, is the corrugation or wrinkling of the crust; and the earth-wrinkles are of three orders of magnitude : continents, mountain-ranges, and rock-folds or arches.

Continents and ocean-basins, although the most important and permanent structural features of the earth's crust, are quite beyond the scope of such a collection as this, except as their principal relief features and their general relations to the earth's crust have been exhibited in the Introductory Collection in the Vestibule.   The forms and distribution of mountain-ranges might be dismissed in the same way; but, unlike continents, the structure of mountains, upon which their reliefs mainly depend, is quite fully exposed to our observation, and is one of the most important fields for the student of structural geology. Mountains, however, as previously explained, combine nearly all the kinds of structure produced by the subterranean agencies, and their consideration, therefore, belongs at the end rather than at the beginning of this section.   The agency of the horizontal compression of the earth's crust in folding, wrinkling, deforming, and breaking the strata has been explained on page 30 and illustrated by suitable models and specimens; and in the printed explanations of the photographs (4) which form the first illustration in this section (9) the main points are stated again.   But our chief purpose now is simply to study in greater detail the various effects thus produced.

Inclined or Folded Strata.—Normally, strata are horizontal, and dikes and veins are vertical or nearly so. Hence the stratified rocks are more exposed to the crumpling action of the tangential pressure in the earth's crust than the eruptive and vein rocks ; and it is for this reason, and partly because the stratified rocks are vastly more abundant than the other kinds, that the effects of the corrugation of the crust are studied chiefly in the former. But it should be understood that folded dikes and veins are not uncommon.

That the stratified rocks have, in many instances, suffered great disturbance subsequent to their deposition, is very evident ; for, while the strata must have been originally approximately plane and horizontal, they are now often curved, or sharply bent and contorted, and highly inclined or even vertical. All inclined beds or strata are portions of great folds or arches. Thus we may feel sure when we see a stratum, as in the model (5), sloping downward into the ground, that its inclination or dip does not continue at the same angle, but that at some moderate depth it gradually changes and the bed rises to the surface again, as is so clearly shown in the printed explanation and section accompanying the model. Similarly, if we look in the opposite direction and think of the bed as sloping upward—we know that the surface of the ground is being constantly lowered by erosion, and consequently that the inclined stratum formerly extended higher than it does now, but not indefinitely higher ; for, in imagination, we see it curving and descending to the level of the present surface again. Hence it forms, at the same time, part of one side of a great concave arch and of a great convex arch, just as every inclined surface on the ground indicates both a hill and a valley. And guided by this principle we can often reconstruct with

much probability folds that have been more or less completely swept away by erosion, or that are buried beyond
our sight in the earth's crust.

The arches of the strata are rarely distinctly indicated in the
topography, but must be studied where the ground has been
partly dissected, as in cliffs, gorges, quarries, etc. They are
also, as a rule, far more irregular and complex than they are
usually conceived or represented. The wrinkles of our cloth-
ing are often better illustrations of rock-folds than the models
and diagrams used for that purpose. This is sufficiently obvious
when we glance at the specimens, or reflect that the earth's
crust is exceedingly heterogeneous in composition and structure,
and must, therefore, yield unequally to the unequal strains im-
posed upon it.

The natural illustrations begin with two comparatively
simple folds or bends of the strata (21-22). The first
one is a complete fold, while the second shows only a
single layer of rock, an outline or skeleton fold. In
their present positions they are typical convex arches or
anticlines, but if inverted they would be equally good
concave arches or synclines; and may thus be used to
illustrate these two principal types of folds.

The imaginary line passing longitudinally through a fold,
about which the strata appear to be bent, is the axis; and the
plane lying midway between the two sides of a fold and
including the axis is the axial-plane. In the anticline (21-22)
the arch is convex upward and the beds slope downwards or
dip on either side away from the axis; while in the syncline (23)
the arch is concave upward and the beds dip toward the axis.
These two types of folds are commonly, but not always, correl-
ative. like hill and valley.

Rock-folds are of all sizes, from almost microscopic wrinkles to great arches miles in length and breadth and thousands of feet in hight. The smaller folds, or such as may be observed in hand-specimens and even in considerable blocks of stone, are commonly called contortions; and it is interesting to notice that they are, in nearly everything except size, precisely like the large folds, so that they answer admirably as geological models and are our main reliance in this part of the illustration, every one of the natural specimens showing more or less typical contortions. Large folds, however, are almost necessarily curves, while contortions, as the specimens show, are frequently angular.

The normal relation of the wrinkles or contortions to the larger folds is very clearly illustrated by the slate from Newton Centre on the second shelf (24). The contortions, it will be observed, are not uniformly distributed over the main fold; and it is obvious on reflection that when the rocks are folded they must be in a state of tension on the convex side of the arch, and in a state of compression on the concave side. Hence the syncline is evidently the normal position for the minor wrinkles, and they are rarely observed elsewhere.

Contortions are also most commonly found in thin-bedded, flexible rocks, such as shales and schists, as most of the specimens show; and when we find them in hard, rigid rocks, like gneiss (41–43) and quartzite (21), it must mean either that the structure was developed with extreme slowness, or that the rock was more flexible then and possibly plastic.

These minor wrinkles or contortions illustrate not only

the principal types of large folds, but also the normal irregularities of folds, showing, among other things, how folds die out (6) and how a single fold may divide into two folds (25) or *vice versa.*

The monoclinal folds, or those that slope in only one direction, the one-sided arches, are seen in the small specimens (1–3) from local ledges, on the upper shelf; and the large specimen (61) in the next section (10) may be regarded as an abrupt and somewhat wrinkled monocline. This type of flexure is evidently related to, and frequently passes into, faults; and it is of great structural importance in the High Plateau region of New Mexico, Arizona, and southern Utah, as may be seen in the relief maps of the Henry Mountains in the west window.

Anticlines and synclines are *symmetrical* when the dip or slope of the strata is the same on both sides and the axial plane is vertical. The great majority of folds, however, as the specimens show, are not only generally irregular, but have at almost all points unsymmetrical cross-sections, the opposite slopes being unequal and the axial plane inclined to the vertical. This means that the compressing or plicating force has been greater from one side than from the other. It must, in the case of anticlines, have acted with the greatest intensity on the side of the gentler slope. When the steep slope approaches the vertical, this tendency is almost unresisted, and when it passes the vertical, gravitation must assist in overturning the fold (25). Such highly unsymmetrical folds, including all cases where the two sides of the fold slope in the same direction, are described as *overturned* or *inverted*, although the latter term is not strictly applicable to the entire fold, but only to the strata composing the under or lee side of it, as shown by one of the folds in the drawing (**1**, section 10),where the strata on either side are numbered in the order of their deposition. On the under side of the anticline the older strata are seen lying conformably upon the newer. This inversion is one of the most important features of folded strata; and it has led to many mistakes in determining their order of succession. In the great mountain-chains, especially, it is exhibited on the grandest scale, great groups of strata being folded over and over each other as we might fold carpets (**2**).

An inverted stratum is like a flattened S or Z, and may be pierced by a vertical shaft three times, as has actually happened in some coal mines. Folds are *open* when the sides are not parallel, and *closed* when they are parallel, the former being represented by a half-open, and the latter by a closed, book. Closed folds, commonly called isoclines, are usually inverted, and when the tops have been removed by erosion (**3**), the repetition of the strata may escape detection, and the thickness of the section be, in consequence, greatly overestimated. Thus, a geologist traversing the section shown in the drawing would see thirty-two strata, all inclined to the left at the same angle, those on the right apparently passing below those on the left, and all forming part of one great fold. The repetition of the strata in reverse order, as indicated by the numbers, and the structure below the surface, show, however, that the section really consists of only four beds involved in a series of four closed folds, the true thickness of the beds in this section being only one eighth as great as the apparent thickness.

Cleavage Structure.—This important structure is now known to be, like rock-folds, a direct result of the great horizontal pressure in the earth's crust.. It is entirely distinct in its nature and origin from crystalline cleavage, and may properly be called lithologic cleavage. It is also essentially unlike stratification and joint-structure. It agrees with stratification in dividing the rock into thin parallel layers, but the cleavage-planes are normally vertical instead of horizontal. And the cleavage-planes differ from joints in running in only one direction, dividing the rock into layers; while joints, as we shall see, traverse the same mass of rock in various directions, dividing it into blocks.

The principal characteristics of lithologic cleavage are illustrated by the specimens. These show, first, that it is limited as a rule to the soft, fine-grained rocks, having its best development in the slates, as witness the roofing slates (**41–42**), school slates, slate black-boards, etc. ; for slate owes its adaptation to these and other uses chiefly to this wonderful structure, commonly called slaty cleavage, which causes it to split or cleave with remarkable regularity into sheets of any desired thinness. An imperfect cleavage structure is sometimes observed in coarser rocks, such as sandstone and conglomerate (**21**), but these are invariably found on examination to be of a soft and slaty character. The practical limitation of the slaty cleavage to the slates is especially obvious where they are interstratified with coarser and harder rocks. Thus the specimen from Nantasket (**23**), which represents a bowlder lying on the shore at the southern end of the beach, shows two layers of gray sandstone separating three layers of dark bluish gray slate. The slate layers exhibit a very distinct cleavage nearly at right angles to the stratification ; but the cleavage planes end abruptly in every instance at the junction with the sandstone and cannot be traced into the latter rock. The large specimen on the third shelf (**44**) embraces two distinct layers, — a layer of soft, pure slate with perfect cleavage, with a broader layer in front of it of a coarser, sandy slate in which the cleavage is quite imperfect and in a somewhat different direction. Where the alternating layers of coarse and fine slate are much thinner, the frequently changing character of the cleavage produces the so-called curly slate (**22**). The best illustration of all, as regards the relations of cleavage-structure to the character of the rocks, is afforded by the other large specimen on the third shelf (**43**). This is chiefly a well-cleaved slate, the cleavage-planes being parallel to the oblique under surface of the specimen. Toward the left end, two oblique lines of stratification are very plainly marked, and parallel with these, through the middle of the specimen, runs a very distinct layer nearly two inches thick of impure limestone. The cleavage does not extend through the limestone ; but it is obvious at the first glance that the force which developed the cleavage in the slate has simply broken, distorted, and faulted the more rigid and unyielding limestone.

The specimens show, second, and with equal clearness, · that the cleavage-planes are usually highly inclined and transverse to the bedding or true stratification. This is easily seen in the large specimens which we have already noticed, and also in some of the others.   In the block of slate **(24)** from Hull the parallel oblique lines mark the stratification ; while the cleavage, which is somewhat indistinct, coincides with the flat surface on which the specimen rests.   The adjoining specimen **(25)** from the great slate quarries of Slatington, Pa., is similar, except that the cleavage is more perfect.   In the variety of roofing slates known as the ribbon slate **(42)** these darker bands marking the stratification can be seen crossing the cleavage surface at various angles.

A third important characteristic of cleavage, but one not so easily illustrated with specimens, is that it is usually associated with folded strata and very commonly with distorted or flattened fossils and nodules.   The slaty pebbles in the cleaved conglomerate **(21)** have been flattened in the plane of cleavage.

Many explanations of this interesting structure have been proposed, but that first advanced by Sharpe may be regarded as fully established.   He said that *slaty cleavage is always due to powerful pressure at right* angles to the planes of cleavage.   All the characteristics of cleavage noted above are in harmony with this theory.   Cleavage is limited to fine-grained or soft rocks, because these alone can be modified internally by pressure, without rupture.   Harder and more rigid rocks may be bent or broken, but they appear insusceptible of minute wrinkling or other change of structure affecting every particle of the mass.   Since the cleavage-planes are normally vertical, the pressure, according to the theory, must be horizontal.   That this horizontal pressure exists and is adequate in direction and amount, is proved by the folds and contortions of the cleaved strata ; for the cleavage-planes coincide with the strike of the foldings, and are thus perpendicular to the pressure horizontally, as well as vertically. The distortion of the fossils in cleaved slates is plainly due to pressure at right angles to the cleavage, for they are compressed or shortened in that direction, and extended or flattened out in the planes of cleavage. Again, Tyndall has shown that the magnetism of cleaved slate proves that it has been powerfully compressed perpendicularly to the cleavage. And, finally, repeated experiments by Sorby and others have proved that a very perfect cleavage may be developed in clay (unconsolidated slate) by compression, the planes of cleavage being at right angles to the line of pressure. When, however, Sharpe's theory had been thus fully demonstrated, the question as to *how* pressure produces cleavage still remained unanswered. Sorby held that clay contains foreign particles with unequal axes, such as mica-scales, etc., and that these are turned by the pressure so as to lie in parallel planes perpendicular to its line of action, thus producing easy splitting or cleavage in those planes. And he proved by experiments that a mixture of clay and mica-scales does behave in this way. But Tyndall showed that the cleavage is more perfect just in proportion as the clay is free from foreign particles, and in such a perfectly homogeneous substance as beeswax, he developed a more perfect cleavage than is possible in clay. His theory, which is now universally accepted, is that the clay itself is composed of grains which are flattened by pressure, the granular structure, with irregular fracture in all directions, changing to a scaly structure with very easy and plane fracture or splitting in one definite direction.   .

Observations on distorted nodules and fossils have shown that when slaty cleavage is developed, the rock is, on the average, reduced in the direction of the pressure to two fifths of its original extent, and correspondingly extended in the vertical direction. Thus, whether rocks yield to the horizontal pressure in the earth's crust, by folding and corrugation, or by the flattening of their con-

stituent particles, they are alike shortened horizontally
and extended vertically; and it is impossible to overesti-
mate the importance of these facts in the formation of
mountains.

Faults or Displacements.—We may readily conceive that the forces which were adequate to elevate, cor-
rugate, and even crush vast masses of solid rock were
also sufficient to crack and break them; and since the
fractures indicate that the strains have been applied
unequally, it will be seen that unequal movements of
the several parts must often result. If this unequal
movement takes place, *i.e.*, if the rocks on opposite
sides of a fracture of the earth's crust do not move together, but slip over each other, a *fault* (**41**) is produced.
The two sides may move in opposite directions, or in the
same direction but unequally, or one side may remain
stationary while the other moves up or down. It is
simply essential that the movement should be unequal in
direction, or amount, or both; that there should be an
actual slip, so that strata that were once continuous no
longer correspond in position, but lie at different levels
on opposite sides of the fracture. The difference in
movement is known as the *throw*, *slip*, or *displacement* of
the fault, and is commonly measured in the vertical direction.

Fault fractures rarely approach the horizontal direction, but are usually highly inclined or approximately
vertical. When the fault is inclined to the vertical, the
actual slipping in the plane of the fault exceeds the vertical throw, for the movement is then partly horizontal,
the beds being pulled apart endwise. The inclination of

faults, as of veins and dikes, should be measured from the vertical and called the *hade*. Faults are sometimes hundreds of miles in length; and the throw may vary from a fraction of an inch to thousands of feet.

A transverse section, such as is shown in the specimen already referred to (**41**) and in most of the specimens and models, does not give the complete plan or idea of a fault; but this is seen more perfectly in the next specimen (**42**). In this little slab of slate we observe an obliquely transverse crack or rift not quite three inches long, and the layers of slate have been elevated on one side of this line and depressed on the other side, producing a fault having a maximum throw of one fourth of an inch. This is a small but very instructive example. We learn from it that a typical fault is a fracture or rift along which the strata have *sagged* or settled down unequally. The most important point to be observed here is that the strata do not drop bodily, but are merely bent, the throw being greatest at the middle of the fault and gradually diminishing toward the ends. In other words, every simple fault must, as in this instance, die out gradually; for we cannot conceive of a fault ending abruptly, except where it turns upon itself, so as to completely enclose a block of the strata, which may drop down bodily; but the fault is then really endless. The specimens on this shelf (third) are nearly all from Slate Island in Boston Harbor. This somewhat inaccessible islet, is probably the most favorable point in the Boston Basin for the study of some of the more important phases of faults; and the series of specimens will repay careful examination. Several of them (**46, 50**), besides the one already referred to, afford complete horizontal plans or views of typical faults. Where two or more faults appear in the same specimen, they are somewhat overlapping, exhibiting the step-like or *en echelon* arrangment characteristic of faults in general, as well as of various other structural features of rocks. In one instance of this kind (50) one of the faults was probably a foot or more in length when entire. But the most interesting of all these specimens is, doubtless, that one (43) in which the fault is cut off squarely and neatly midway of its length, thus combining the plan and section and giving a better idea of what a fault really is than pages of description.

The beautifully banded sandstones from the vicinity of the Hot Springs, in the Black Hills of South Dakota, (62-65) are also excellent illustrations, on a small scale, of various types of faulting. It will be most convenient, however, and save repetition, to present the different phases of faults systematically and refer to the specimens, so far as they are illustrative, in the same order.

The rock above an inclined fault, vein, or dike is called the *hanging wall*, and that below the *foot wall*; and inclined faults are divided into two classes, according to the relative movements of the two walls. Usually, the hanging wall slips down and the foot wall slips up. Faults on this plan are so nearly the universal rule that they are called *normal* faults. They indicate that the strata were in a state of tension, for their broken ends are pulled apart horizontally, so that a vertical line may cross the plane of a stratum without touching it.

A few important faults have been observed, however, in which the foot-wall has fallen and the hanging-wall has risen.

These are known as *reversed* faults; and they indicate that the strata were in a state of lateral compression, the broken ends of the beds having been pushed horizontally past each other, so that a vertical line or shaft may intersect the same bed twice, as has been actually demonstrated in the case of some beds of coal.

All of the models and nearly all the specimens show normal faults, the only clear exception being one of the Dakota specimens (**62**). This shows no fewer than five distinct and approximately parallel displacements, and in each case the hanging wall has risen relatively to the foot wall, the broken ends of the layers of sandstone overlapping. Obviously, this means horizontal compression and consequent slipping along a series of oblique fractures. The usual explanation of normal faults is that when the rocks are in a state of tension, a tendency exists to widen the fractures, and the hanging wall being thus left unsupported naturally drops down. The sufficiency of this explanation is especially obvious in the case of converging fractures, enclosing large, V-shaped blocks of strata. When these great wedges settle down the bounding fractures become normal faults. Several examples of these converging faults, which are also called trough-faults and compensating faults, are shown in the models (**2, 21**). Trough-faults, it may be added, are often limited in depth, since the displacement does not necessarily extend below the V-shaped block. A still better explanation, recently proposed by Professor Le Conte, introduces the principle of flotation and is illustrated by the blocks of wood on the back part of the second shelf (**24**). These blocks, it will be observed, were made by cutting a piece of plank obliquely the requisite number of times; and the plank thus divided may be compared to a portion of the earth's crust traversed by oblique fractures. Now imagine the blocks, all in their original positions, as floating on water and free to adjust themselves in obedience to the force of gravity. It is obvious that they will not retain their original horizontal positions, but each block will be tilted, and independently of the others, the over-hanging or heavy end settling down and the light end rising, until it gains the position of equilibrium shown in the present arrangement of the blocks, each block having been tested by floating it on water before placing it on the shelf. The point of special interest, of course, is that, after the principle of flotation has thus asserted itself, we find that each cut or plane of division in the plank has become a fault, and a normal fault, the hanging wall, in every instance, having risen and the foot wall settled down; and the earth's crust must, of course, tend to behave in a similar manner, if we, as the facts warrant, regard it as a layer divided by oblique fractures and resting upon a relatively mobile substratum or foundation.

Important reversed faults are believed to occur chiefly along the axes of overturned anticlines, where the strata have been broken by the unequal strains, and those on the upper side shoved bodily over those on the lower or inverted side; but, independently of this explanation, folds and faults are closely related phenomena. In the former the strata are disturbed and displaced by bending, in the latter by breaking and slipping; and the displacement which is at one point accomplished by a fold may, as the displacement increases or the rocks become more rigid, gradually change to a fracture and slip. This relation is especially noticeable with monoclinal folds, in which the tendency to break or shear the beds is often very marked. The large relief maps in the window cases of this room embrace clear examples on a large scale of closely related and connected faults and monoclines; and on a small scale we have the Slate Island faults on the second shelf, each one of which may be regarded as a modified monocline, the flexing having been sharp enough to overcome the cohesion of the slate at some points but not at others. In two of these specimens (**44, 48**) there is no actual faulting, but the tendency to shear the slate is seen in the fact that it is pinched to about half its normal thickness along the axis of the fold. Again, in consequence of this pinching, the faults may die out vertically as well as horizontally. Thus, in another of the Slate Island specimens (**46**) the well-marked fault which we see on the upper surface is represented on the lower surface only by a faintly marked monocline.

Faults cutting inclined or folded strata are divided into two classes, according as they are approximately parallel with the direction of the dip or of the strike. The first are known as *transverse* or *dip* faults, and the second as *longitudinal* or *strike* faults. The chief interest of either class consists in their effect upon the outcrops of the faulted strata, after erosion has removed the escarpment produced by the dislocation. Dip faults cause a lateral shift or displacement of the outcrops, as is clearly shown by the north-south fault in the plaster model (**23**), the fault being recognized in this case not only by the shifting of the bands of color marking the outcrops of successive strata, but also of the topographic features or ridges due to the unequal erosion of the strata. By reference to the direction of dip of the strata it is easily seen that the down throw side of this fault is on the right and the upthrow on the left. If the throw of the fault were reversed, the displacement of the outcrops would be reversed also. This model also shows a very clear strike fault. The effect of this fault is, evidently, to repeat the outcrops of the strata, the beds south of the fault being identical, as shown by the colors, with those north of it. This repetition of the strata in the same order proves the existence of a fault, as explained on the label, since repetition by folding is necessarily in the reverse order. Although strike faults thus commonly repeat the strata, and cause the apparent aggregate thickness of beds in the section to exceed the real thickness, they sometimes have the opposite effect, concealing a portion of the beds and making the apparent thickness less than the real thickness.

An extensive displacement of the strata is sometimes accomplished by a short slip along each of a series of parallel fractures, producing a step fault, as is so clearly shown in two of the Black Hills specimens (62–63), the fault being reversed in the first and normal in the second; and the normal series shows particularly well how a step faults simulates, in the general view, a monocline. Important faults are rarely simple, well-defined fractures; but, in consequence of the enormous friction, the rocks are usually more or less broken or crushed, sometimes

for a breadth of many feet or yards. This feature, although especially characteristic of large faults, is well illustrated by one of the Black Hills specimens **(65)**.
This is crossed obliquely by a very distinct fault; and
along either side of the fault the rock for a total breadth
of about an inch is finely sheeted by numerous cracks
approximately parallel with the main fracture. One
result of this comminution is that fragments of the various beds are often strung along the fault in the opposite
direction to the slipping, and this circumstance has
been made use of in tracing the continuation of faulted
beds of coal. In other cases the direction of the slip is
plainly indicated by the bending of the broken ends of
the strata, and the beds are sometimes turned up at a
high angle or even overturned in this way. This is
clearly shown in another of the Black Hills specimens
(64), the layers along the very strongly marked oblique
fault being flexed upward on the downthrow side ; but,
on account of the local bleaching of the colors, the downward flexing on the upthrow side can not be seen.

Since fault fractures are not usually plane, but undulating
and often highly irregular, the walls will not coincide after
slipping; and if the rocks are hard enough to resist the enormous pressure, the cavities or fissures produced in this way
may remain open. Now faults are, in many cases, continuous
fractures of the earth's crust, reaching down to unknown but
very great depths and affording outlets for the heated subterranean waters; so that it is common to find a fault marked on
the surface by a line of springs, and these are often thermal.
The warm mineral waters on their way to the surface deposit
part of the dissolved minerals in the irregular fissures along the
fault, which are thus changed to mineral veins. This agrees

with the fact that the walls of veins, as we have already seen, (page 232), usually show faulting, as well as crushed rock, slickensides, and other evidences of slipping.

The colored layers coinciding with some of the faults in the wooden model (22) are designed to represent mineral veins ; still better, in several of the Slate Island specimens (47, 49) the fault fissures are occupied by veinlets of calcite,— small but exceedingly instructive examples. In one of these (45) the slate is much broken ; and, although the calcite has been nearly all dissolved out, it may be regarded as, in plan at least, a miniature stockwork (page 241). The same general relations are observed, but less commonly, with dikes ; and we have a superb illustration in the large specimen from Ontario (81). Crossing the bedded diorite are two nearly vertical dikes of granite, and along the right hand dike especially the diorite is distinctly faulted, the downthrow being on the left and nearly two inches, as shown by the feldspathic layers. The two specimens on the same shelf from Marblehead (82-83) are also very interesting in this connection. The first one is a mass of diorite traversed by a very regular dike of fine grained syenite ; and this is sharply faulted twice by two later, oblique, and very slender dikes of the same material, the right hand slip being about two and a half inches in the plane of the fault. The second specimen, on the other hand, shows two parallel dikes of syenite in diorite broken nearly at right angles and slipped about two inches by a single prominent dike of syenite, the downthrow being on the left.

The faulted pebble of quartzite (61) is from the Roxbury pudding stone. The remarkably regular joint-planes
of that rock frequently divide the pebbles ; and it is a
common observation that the two parts of the same pebble no longer correspond in position, but show a slip or
displacement in the plane of the fracture.

The large piece of interstratified slate and sandstone
from Brighton (66) shows one distinct and several indistinct faults, all characterized by a normal degree of
irregularity. But the small dike in white marble from
Smithfield, R. I. (67), has been dislocated beyond any
possible restoration. The numerous dikes of the Boston
Basin afford many admirable illustrations cf different
kinds of faults ; but undoubtedly the best single example
is the dike on the Swampscott Shore which is represented,
on a scale of twelve feet to the inch, in the drawing (24).
It is perfectly exposed for a length of nearly four hundred feet, and is broken in that distance by no fewer
than thirty-six distinct faults.   .

The wooden models on the first two shelves illustrate
some of the more general relations of faults and of systems of faults, and especially the relations of faults to
erosion. The ages of faults are determined in the same
ways as the ages of dikes. They are, of course, always
newer than the rocks which they intersect ; and where
faults of different systems or directions intersect, the
newer usually displaces the older. One of the models
(22) shows this particularly well, the northwest-south-
east fault being clearly newer than the northeast-south-
west fault.

If the earth's surface were not subject to erosion, nearly

every fault of any magnitude would be marked on the surface
by an escarpment equal in hight to the throw of the fault, as
shown in several of the models (21-22); and, notwithstanding
the powerful tendency of erosion to obliterate them, these escarpments are sometimes observed, although of diminished
hight. Thus, according to Gilbert, the Zandia Mountains in
New Mexico are due to a fault of 11,000 feet, leaving an escarpment still 7,000 feet high. The relief maps of the Henry Mountains in the west window show several such fault-scarps, and
they are a prominent feature of the relief map of the Grand
Cañon district in the south window. But, as a rule, there is no
escarpment or marked inequality of the surface, faults, like
folds, not being distinctly indicated in the topography. In all
such cases we must conclude either that the faults were made a
very long time ago, or that they have been formed with extreme slowness, so slowly that erosion has kept pace with the
displacement, the escarpments being worn away as fast as
formed. These and other considerations make it quite certain
that extensive displacements are not produced suddenly, but
either grow by a slow, creeping motion, or by small slips many
times repeated at long intervals of time.

The relations of faults to erosion and their appearances on
geological maps are farther illustrated by the wooden models,
but the special features of each are explained on the labels, and
need not be referred to here.

Joints and Joint-structure. — This is the most
universal of all rock-structures, since all hard rocks and
many imperfectly consolidated kinds, like clay (3), are
jointed. Joints are cracks or planes of division which
are usually approximately vertical and traverse the same
mass of rock in several different directions. They are
distinguished from stratification planes by being rarely
horizontal, and from both stratification and cleavage
planes by being actual cracks or fractures, and by divid-

ing the rock into blocks instead of sheets or layers. The art of quarrying consists in removing these natural blocks; and most of the broad, flat surfaces of rock exposed in quarries, are the joint-planes (4). Some of the most familiar features of rock-scenery are also due to this structure,—cliffs, ravines, etc., being largely determined in form and direction by the principal systems of joints; and we have already seen that the same is true of veins, dikes, and faults.

Joints are divided by their characteristics and modes of origin into three classes as follows:—

1. *The parallel and intersecting joints.* This is by far the most important class, and has its best development in stratified rocks, such as sandstone, slate, limestone, etc., and especially in the fine grained, brittle rocks. These joints are straight and continuous cracks which may often be traced for considerable distances on the surface. They usually run in several definite directions, being arranged in sets or systems by their parallelism. Thus, in the drawing (4) one set of joints is represented by the broad, flat surfaces in light, and a second set crossing the first nearly at right angles, by the narrower faces in shadow. One of the specimens on the second shelf (21) is an exceptionally good illustration of a single system of joints. Including the ends, it shows five even and closely parallel breaks or joint-planes. By the intersection of the different sets of joints the rock is divided into angular blocks, the forms of which depend upon the regularity, and angles of intersection of the joints. The more usual or normal forms of the joint-blocks are shown in the specimens, largely from the slate

quarries of this vicinity, on the second shelf. Most of
the blocks are bounded by two systems of joints and the
bedding planes, the latter being horizontal, or forming
the upper and under surfaces of the blocks as they are
now placed. Occasionally the intersections of the joints
are approximately rectangular, yielding cuboidal blocks
(**22**), but more commonly they are oblique and the
blocks are rhombic in form (**25, 27, 30**). In these
cases the joints may or may not be oblique to the bedding planes. In one instance (**23**) the bedding surface
is ripple-marked. Several of the blocks (**26, 28**) are
bounded by joints of three different systems, and those
bounded by four or five systems are not uncommon in
the ledges. The two rhomboidal blocks of trap (**29, 32**)
are interesting simply because the forms are unusually
regular for rock of that character; and the jointing of
the Roxbury pudding stone (**31**) is wonderfully regular
and perfect, when we consider what a coarse, hard, uneven stone it really is. The two large blocks of slate on
the third shelf (**41–42**) are intended especially to show
how remarkably smooth and even the joint-faces sometimes are in rocks of that character.

Joints not only facilitate mining and quarrying operations, but
they also determine the size and character of the stones which a
quarry can produce. Thus most of the slate around Boston is
unsuited for building purposes, because it is too finely jointed.
Material suitable for monolithic columns can, obviously, only
be obtained where the joints are far apart in at least one direction. This condition is realized in miniature in the specimens
of slate from Hult's Cove (**1**), the joint-blocks being distinctly
prismatic in form. The long specimen on the back part of the

second shelf (34) is rather exceptional in its dimensions. It was originally bounded by two parallel joint planes on the front and back edges, and by bedding planes above and below, but erosion has nearly effaced all of these. When the joints of one system are very close together the rock is sheeted and the jointing resembles stratification. The slab of granite from a quarry in Hingham (68) is an excellent illustration. This sheet-jointing is distinguished from that of the third class by not being horizontal. Joints are usually approximately plane, and the marked curvature seen in one of the specimens (2) is quite unusual. The jointed Miocene clay (3) is really semi-lithified; but joints are not uncommon in the plastic glacial clays of New England.

Although many explanations of this class of joints have been proposed, it has long been the general opinion of geologists that they are due to the contraction of the rocks, *i. e.*, that they are shrinkage cracks. We shall soon see, however, that they lack the most essential characters of cracks known to be due to shrinkage. More recently Daubrée has proposed to regard them as due to torsional strains, to the bending and twisting of the rocks; and this is undoubtedly a true explanation. The strains would, however, be developed too slowly to explain the remarkably regular fractures often observed, especially in rocks of coarse and irregular texture like the Roxbury pudding stone. To meet this difficulty, the present writer has advanced the view that the swift vibratory movements of the rocks known as earthquakes are an important if not principal cause of parallel joints. It is well known that earthquakes break the rocks; and it can easily be shown that the earthquake fractures must possess all the essential features of parallel and intersecting joints.

2. *The contraction joints or shrinkage cracks.* That many cracks in rocks are due to shrinkage, there can be no doubt. The shrinkage may result from the drying of sedimentary rocks; but more generally from the cooling

of eruptive rocks. Every one has noticed in warm weather the cracks in layers of mud or clay on the shore, or where pools of water have dried up (65) ; and we have already seen that these sun-cracks are often preserved in the hard rocks (66). The true shrinkage cracks have certain characteristic features by which they may be distinguished from the joints of the first class. They divide the clay into irregular, polygonal blocks, which often show a tendency to be hexagonal rather than quadrangular. The cracks are continually uniting and dividing, but are not parallel, and rarely cross each other. Sun-cracks never affect more than a few feet in thickness of clay, aud are an insignificant structural feature of sedimentary rocks. In eruptive rocks, on the other hand, the contraction joints have a very extensive, and, in some cases, a very perfect development, culminating in the prismatic or columnar jointing of the basaltic rocks. This remarkable structure has long excited the interest of geologists, and, although the basalt columns were once regarded as crystals, and later as a species of concretionary structure, it is now generally recognized as the normal result of slow cooling in a homogeneous, brittle mass. The columns are normally hexagonal, and perpendicular to the cooling surface, being vertical in horizontal sheets and lava flows, as in the classic examples of the Giant's Causeway (61) and Fingal's Cave, and horizontal in vertical dikes (67). Vertical columns are sometimes called "palisades," as on the west bank of the Hudson above New York City ; and "the devil's organ" is another common name. The columns begin to form on the cooling surface of the mass, and gradually extend toward the center, so that dikes sometimes show two independent sets of columns. This transverse prismatic or columnar jointing is imperfectly developed in many of the trap dikes about Boston; and at one point in Needham vertical columns are well developed in a surface flow of felsite. The small columns of basalt (82) are from the volcanic district of the Rhine, representing material that is extensively quarried for the manufacture of paving stones. The Giant's Causeway is represented by the model (61) and also by portions of two columns, one in the case (81) and one in the right window space. The columns are commonly divided at short intervals, as in these examples, by curving transverse fractures, which are attributed to the unequal cooling of the columns. The column of argillaceous limestone from the Delaware Water Gap, Pa., (83) represents a stratum in which prismatic jointing due, apparently, to shrinkage is well developed.

Shrinkage cracks in sedimentary deposits may be the result, in some cases, of volcanic heat instead of solar heat. Thus the columnar specimens of slate (62) and sandstone (63–64) on the fourth shelf are from the walls of dikes and might be classed among the contact phenomena. When the dikes were formed, the volcanic heat escaped from the trap into the sandstone and slate; and while the columnar structure was slowly developed in the trap by cooling, it was developed with almost equal regularity, but on a small scale, in the wall rocks, by heating and consequent desiccation, the columns being, in both cases, perpendicular to the walls.

3. *The concentric joints of granitic rocks.* In quarries of granite and other massive crystalline rocks, it is often

very noticeable that the rock is divided into more or less
regular layers by cracks which are approximately parallel
with the surface of the ground (54), some of the granite
hills having thus a structure resembling that of an onion.
The layers are thin near the surface, become thicker and
less distinct downward, and cannot usually be traced below a depth of fifty or sixty feet.   These horizontal
cracks are of great assistance in quarrying, and are now
regarded as due to the expansion of the superficial portions of the granite under the influence of the solar heat.
In reference to this view of their origin these may be
called *expansion joints*.

  Slickensides and Stylolites.   These are minor phenomena related to joint-structure and illustrated by the
specimens on the third shelf.   Slickensides is the name
given to the polished and striated surfaces often observed
on joint-planes and on the walls of veins and dikes.   This
appearance is commonly explained as due to the friction
when the rock surfaces were slipping over each other
under great pressure ; and this mechanical theory is probably applicable where the slickensided surfaces are of the
same nature as the mass of the rock, as in the case of the
polished iron ore (43), coal (48), and quartz (49).   In
other cases, however, the slickensides exist only in a
special mineral deposit on the rock surfaces, such as
chlorite (44), epidote (47), iron oxide (50), and pyrite
(45); and we may suppose that slipping occurred during
the deposition of the minerals ; although with chlorite
(44), serpentine (57) and other hydrous species the
mechanical explanation may be dispensed with and the
slickensides referred to the swelling and consequent

crowding and squeezing attending the development of the minerals from the rocks which they cover.

Stylolites are related to slickensides, although not connected with joint-planes. They may usually be described as vertical surfaces or cracks in fine grained rocks like slate and limestone (51-53) which may enclose a distinct pencil or column of the rock (53) or wind indefinitely through the rock, like a fluted curtain (51-52). The German specimen (51) is a particularly good example of this type. The structure is to be explained in most cases at least as due to a differential vertical slipping initiated, perhaps, by the weight of overlying strata. The two examples of a wavy or undulating fracture (55-56) illustrate the breaking of homogeneous rocks when in a state of vibration; and although such artificial fractures are distinct from jointing, they may be most conveniently classified with them.

Concretions and Concretionary Structure. — Folds, cleavage, faults, and joints — all the subsequent structures considered up to this point — are the product of mechanical forces. Chemical agencies, although very efficient in altering the composition and texture of rocks, are almost powerless as regards the development of rock structures; and concretions are the only important structural feature of rocks belonging in this class and having a chemical origin. Although of minor importance, this is one of the prettiest and most interesting divisions of structural geology, and the illustrations are, fortunately, unusually complete and satisfactory.

Concretions are formed by the segregation of one or

more of the constituents of a rock. But there are three
distinct kinds of segregation. If the water percolating
through or pervading a rock, dissolves a certain mineral
and afterwards deposits it in cavities or fissures, *amygdules, geodes,* or *veins* are the result. If the mineral is
deposited about particular points in the mass of the rock,
it may form *crystals*, the rock becoming *porphyritic;* or it
may not crystallize, but build up instead the rounded
forms called *concretions*, the texture or structure of the
rock becoming *concretionary*. A great variety of minerals occur in the form of concretions, but this mode of occurrence is especially characteristic of certain constituents
of rocks, such as carbonate of lime, carbonate of iron,
oxide of iron, and silica. Concretions may be classified
according to the nature of the segregating minerals ; and
in each class we may distinguish the *pure* from the *impure* concretions. A pure concretion is one entirely
composed of the segregating mineral. Most nodules of
flint and chert, quartz geodes, concretions of pyrite,
and many hollow iron-balls are good illustrations of this
class. In all these cases the segregating mineral has
been able in some way to remove the other constituents
of the rock, and make room for itself. But in other
cases it has lacked this power, and has been deposited
between and around the grains of sand, clay, etc. ; and
the concretions are consequently impure, being composed
partly of the segregating mineral, and partly of the other
constituents of the rock. The calcareous concretions
known as clay-stones are a good example of this class,
being simply discs of clay, all the minute interstices of
which have been filled with segregated calcite. The

solid iron balls are masses of sand filled in a similar manner with iron oxides.

Concretions are of all sizes, from those of microscopic smallness in some oölitic limestones up to those twentyfive feet or more in diameter in some sandstones.

The point of deposition, when a concretion begins to grow, is often determined by some concrete particle, as a grain or crystal of the same or a different mineral, a fragment of a shell, or a bit of vegetation, which thus becomes the nucleus of the concretion. The ideal or typical concretion is spherical; but the form is influenced largely by the structure of the rock. In porous rocks, like sandstone, they are frequently very perfect spheres; but in impervious rocks like clay, they are flat or disc-shaped, because the water passes much more freely in the direction of the bedding than across it; while the concretions in limestones, the nodules of flint and chert, are often remarkable for the irregularity of their forms. In all sedimentary rocks the concretions are arranged more or less distinctly in layers parallel with the stratification, which usually passes undisturbed through the impure concretions. Many siliceous and ferruginous concretions are hollow, apparently in consequence of the contraction of the substance after its segregation; and the shrinkage due to drying is still further indicated by the cracks in the septaria stones. The hollow siliceous concretions are usually lined with crystals (geodes), while the hollow iron-balls frequently enclose a smaller concretion. Rocks often have a concretionary structure when there are no distinct or separable concretions. And the appearance of a concretionary structure (pseudo-

concretions) is often the result of the concentric decomposition of the rocks by weathering.

With this general introduction to the concretions,
which will enable us to avoid needless repetition, we may
proceed to notice the specimens more in detail.

The illustration begins ·with a very complete series of the
impure calcareous concretions or clay-stones, chiefly from the
clay beds of the Connecticut Valley. A small proportion of
carbonate of lime or calcite, which must once have been diffused through the whole mass of the clay, has subsequently undergone segregation; and being deposited between the particles
of clay, has cemented them together to form the hard, round,
concretions or clay-stones, which are now composed in nearly
equal parts of clay and the segregating mineral—calcite. Although the normal form is evidently discoid, the actual forms
are almost infinitely varied, but always graceful and frequently
exquisitely beautiful. They have, more than most concretions,
a strangely artificial appearance; and it is hard at first to believe that some of them are not the product of the potter's
wheel or the turning lathe. In two of the specimens (1, 12)
the discoid form is carried to an extreme, the inference being
that in these cases the clay beds were very impervious to water
in directions across the bedding, so that the concretions could
only grow in the planes of stratification. The elongated forms
(26, etc.) indicate movement of water in a definite direction in the
plane of stratification; and the composite forms (9, etc.) mean,
of course, that what were originally separate concretions became united by their continued growth. The annulation or
concentric rings characteristic of many of the clay stones is
due partly to intermittent growth and partly to ·the freer passage of water along certain laminae of clay than others. The
smaller clay-stones show, as we should naturally expect, a
nearer approach to a spherical form. But as soon as the vertical diameter equals the thickness of the pervious layer of clay,
the growth becomes almost wholly horizontal. A visible nu-

cleus is commonly wanting in clay-stones; and the specimen in which the segregation has evidently taken place around a calcareous shell (5) is, therefore, somewhat exceptional. Sections of clay stones, when polished and slightly etched, often show faint concentric lines (4), proving that the growth has not been perfectly uniform. The sections also show the lines of stratification passing undisturbed through the concretion.

The sandstone specimen (18) differs from the claystones in being more spherical, a necessary result of the more permeable character of sand. The lines of stratification are plainly marked on the exterior, and the general elongation indicates a current of water flowing obliquely across the strata in the direction of the smaller ends of the concretions. Pure calcareous concretions are practically unknown, except in the case of the minute forms which we have in pisolite (10) and oölite (11). These are varieties of limestone in which the calcite is largely in the form of minute spherical concretions; and these cannot usually, if ever, be regarded as subsequent structures. The segregating mineral in these cases is not, as usual, a rare constituent, but the chief constituent, of the rock; and the carbonate of lime has in no case power to remove the substance of the rock and make a clear space for itself. The conclusion follows, therefore, that these little spheres of pure calcite must be formed when the rock itself is made. Fortunately, oölitic and pisolitic limestones are now forming in many places; and, as the specimen from Great Salt Lake (15) shows, we have at first a concretionary sand; each minute grain is a spherical concretion before the consolidation of the rock begins. Microscopic examination shows that each of these little spheres has as a nucleus an angular grain of calcite; and the mode of formation is simply this: as the fine calcareous sand due to the comminution of shells and corals is drifted about by the waves, each minute grain is slowly incrusted by thin concentric layers of the carbonate of lime which the sea-water holds in solution, and with the subsequent cementation of these enlarged and rounded grains the oölitic limestone is complete. All this is shown very clearly in the magnified section of oölite in the lithological col-collection (83 section 14).

The silicious concretions occur chiefly in calcareous rocks and are usually pure, the segregating silica having power to remove the carbonate of lime and make a free space for itself. The segregated silica of the older limestones is known chiefly as chert and hornstone (29–32) ; of the chalk and newer limestones as flint (27–28) or menilite, the latter being a form of hydrated silica or opal. Silicious concretions may also be distinguished as solid, ordinary flint and chert nodules (27–32), and hollow, geodes (33, 81–82). When small, the solid nodules are often spherical (27) ; but with increased size they become, as a rule, highly irregular (31, 35). The usual explanation of the quartz geodes is that into a spherical cavity formed by solution in a massive limestone the silica is introduced in solution or in a gelatinous condition, forming subsequently, as it gradually deposits or dries up, a hard layer on the wall of the cavity and leaving the central part hollow. The silica first deposited is usually compact or chalcedonic, but the last layer is often crystalline (81). The solid concretions grow from the center outward, and the hollow concretions or geodes from the circumference inward. In either case the growth is by concentric layers, and these can be seen very distinctly in one of the specimens of chert (36). The permeability of the solid shell of the geode by water is pretty well proved by the water geodes, and still more clearly by the geodes enclosing bitumen (60). It is likely that these were once ordinary quartz geodes, that the enclosing strata became saturated with petroleum, which, under great pressure, penetrated the shell of the geode and accumulated in its exterior, where it has subsequently undergone partial inspissation or completely dried up, yielding in one case a tarry substance or liquid pitch, and in the other a solid asphalt.

The ferruginous concretions, or those in which iron oxide is the segregating substance, occur very largely in arenaceous rocks. The iron oxide, during its segregation, is unable to remove the sand, and hence they are usually quite impure, being

In most cases simply solid balls of sand cemented by iron oxide. The purer forms, however, are usually hollow. In consequence of the porous nature of the enclosing rock, the iron-balls grow, as a rule, with equal facility in different directions and are commonly globular in form, the smaller examples (41-42) being often very perfect spheres. The fine series from Kentucky (61-64) show how this normal form may be modified, as previously noticed with the calcareous concretions (18), by the movement of the water in a definite direction. In the case of the hollow iron-balls (third shelf), it is necessary to suppose that the iron oxide first segregated or accumulated in a gelatinous condition, and as the subsequent desiccation necessarily began on the exterior of the mass, a hard outer shell was thus formed which increased in thickness by additions to the interior, until the enclosed iron oxide was all used up, and there was left in the central part of the concretion only the original sand, quite free from iron oxide. The normal hollow iron-ball is, then, a solid shell of sand cemented by iron oxide, and enclosing a mass of loose sand. The sand still remains in one of the broken specimens (41) but it has been poured out of the others. The cylindrical concretions ("pipe ore") (60^b-60^c) require no additional explanation, save that the structure of the rock has determined an elongated instead of a globular form. In some cases (49), especially where the rock is of a compact and somewhat impervious character, the iron oxide, during the hardening of the concretion, does not add itself wholly to the outer shell, but the last portion forms a solid and separate nucleus,—a concretion within a concretion, the kernel of the nut or yolk of the egg, concretions of this kind, especially, being very commonly mistaken for fossil nuts, fruits, etc. Most curious of all are the annular concretions or hollow iron rings (45). These are formed where the gelatinous iron oxide, segregating in a pervious layer of limited thickness, assumes first a discoid form, which, as it hardens and shrinks, becomes a semi-gelatinous ring; and this, like the globular masses, becomes hollow by more complete desiccation. The angular specimens

(60ᵃ) show how the concretionary structure may be modified by the pre-existing joint-structure of the rock, the segregation of iron oxide taking place in the periphery of the joint-blocks.

From the concretionary forms of iron oxide we pass naturally to those of iron sulphide—pyrite and marcasite. These are commonly globular or discoid according to the pervious or impervious character of the enclosing rock, and are always solid. The large specimens from Newfoundland (68-74) show a good gradation in form and in the coarseness and distinctness of the crystallization, some of them being also composite forms (71-73).

Undoubtedly the most perfect and interesting specimens in this group are the large sphere (67) and ring (66). These are solid pyrite and almost unique in their ideal symmetry of form. The nodules of chalcocite (57) show that this form is not limited to iron sulphide. In this connection also we may notice the concretions of feldspar in spherulitic obsidian (55) and felsite (54); of mica (58) in the celebrated granite of Craftsbury, Vermont; of hornblende (59) and of ulexite (56).

Carbonate of iron (siderite) is commonly associated in nature with carbonate of lime (calcite), and nowhere more commonly or intimately than in concretions. It is the carbonate of iron which causes the claystones to weather brownish (9-12); and it is obviously present in most of the specimens on the lower shelf. The latter are especially interesting, however, as examples of septaria-stones. These are due to the segregation of the carbonates of lime and iron in clay, the hardening of the exterior, and the development of shrinkage cracks in the interior, which subsequently become veinlets of calcite and siderite. Four of the specimens showing the septaria structure are artificial sections (83-84, 87-88), while the others are concretions which have been partly worn away by natural erosion.

The illustrations of concretionary structure begin with instances of the segregation of the coloring matter of rocks, as in the dendrite (1-2, 6), where we have branching stains of iron and manganese oxides, chiefly on the joint and bedding surfaces of the rocks, and in the leopardite (3-5), where the stains tend to be cylindrical and penetrate the unbroken stone. The remaining specimens on this shelf (7-9) are instances where the segregation of carbonate of lime in beds of sand has developed concretionary shapes but no distinct or separate concretious. In two of the specimens on the next shelf (22-23) the concretions are somewhat distinct, but still connected. The curiously reticulated forms on this shelf (21-25) are less easily explained, but we seem to have a clue to the correct explanation in the ragged edge of the smaller one; irregularly branching processes gradually interlace, forming holes which are subsequently filled up by concentric deposits.

The curiously and regularly dimpled plates of argillaceous limestone (24) are especially puzzling. The lines of deposit in the limestone conform with the curves; and it is at least doubtful if, strictly speaking, these are concretions at all. The beautiful and closely crowded dimples on the weathered surface of marble (26) must, however, be regarded as the external development by delicate erosive action of a true concretionary structure, although no trace of this can be seen on the fresh fracture of the marble.

The silicified corals from Florida (27-28) may be regarded as true chalcedony geodes whose forms have been determined by the solution of heads or masses of coral in beds of marl; in other words, they are segregations of silica pseudomorphic after coral. And with the same explanation we may dismiss the silicified wood (29-30). Undoubtedly few of the structures classed under the general head of concretions are more curious than the cone-in-cone (41-48). The name is descriptive, the structure consisting of corrugated or crenulated conical layers, one within another, and in the more complete specimens (41, 47) it is seen that thin layers of the rock, a calcareous and sideritic clay, is composed of the closely crowded nests of cones, the axes of the cones being transverse to the bedding planes. The hight of the cones measures the thickness of the layers, which is commonly one to four inches, exceptionally six inches or more (48). It seems necessary to suppose that during the compression of the layer of clay by vertical pressure, it is divided by an indefinite series of conical gliding surfaces, which are corrugated by the intermittent character of the movement. One of the specimens (47) appears to be part of a large lenticular concretion, the outer layers of which show the cone-in-cone structure; and in another specimen (46) the structure is cylindrical rather than conical.

The other clay specimens on this shelf (52–54) are rudely concretionary forms developed in clay, as those on the upper shelf were in sand. The large, rounded and cylindrical, hollow concretions of iron oxide on this shelf and the next require no explanation beyond what has been given.

What may not inaptly be called pseudo-concretions are seen in the concentrically weathered joint-blocks of diabase (67–69), the concretionary form showing most distinctly where the decomposed material has been exfoliated (67). The three large specimens on the bottom shelf (81–83) show the natural end, transverse section and longitudinal section of what has been regarded as a cylindical type of concretion, but is now known to be a partly silicified and partly calcified palm tree; and it belongs here only in the same sense as the ordinary silicified wood on the second shelf.

SUBSEQUENT STRUCTURES PRODUCED BY THE SUPERFICIAL OR AQUEOUS AGENCIES.

The superficial agencies, as we have seen, are, in general terms, water, air, and organic matter. Geologically considered, the results which they accomplish may be summed up under the two heads of deposition and erosion—the formation of new rocks in the sea, and the destruction of old rocks on the land. In the role of rock-makers they produce the very important original structures of the stratified rocks; while as agents of erosion they develop the most salient of the subsequent structures of the earth's crust—the infinitely varied relief of its surface. As a general rule, to which recent volcanoes are one important exception, the original and subterranean structures of rocks are only indirectly, and often very slightly, represented in the topography; for this, as we have seen, is almost wholly the product of erosion. Therefore, what we have chiefly to consider are the more characteristic forms produced by erosion, and, incidentally, the extent to which the erosive action is influenced by the pre-existing structures of the rocks.

This subject, or department of petrology, presents two quite distinct phases, according as we take a general or detailed view of it. The first is virtually a study of topography in its geological relations; while the second takes account of those minor but highly significant erosion-forms which have little or no topographic or scenic value. The first or major phase is ill-adapted to Museum illustration, for we must depend wholly upon pictures and models; and as these illustrations are somewhat scattered about the room, we may most conveniently take up this section last.

The minor erosion-forms can be illustrated to a considerable extent by natural specimens; and these may be most naturally classified in accordance with the general agencies concerned in their development.

Minor Erosion Features.

Chemical Erosion.—Among the more characteristic forms due to the chemical erosion or decay of rocks,

aided, in many cases, by rain and frost, are : (1) The
concentrically decomposed and exfoliated joint-blocks
(21-22) of trap and other silicate rocks, of which the
dikes of this vicinity afford admirable illustrations. (2)
The hard, irregularly eroded or etched masses of similar
rocks (82-83) which have been exposed to the erosive
waters of bogs and marshes. Such examples are common in the bogs and meadows of this region, and in their
fantastic shapes testify to the solvent power of the organic acids, the less readily decomposed parts being left
in relief. (3) The cellular structure due to the chemical
decay of crystals of pyrite (1) and other easily decomposed minerals, and to the solution of fossils (2). (4)
Surfaces roughened by crystals (3), concretions (4), or
silicified fossils (5) being left in relief through the re-
moval of the enclosing rock. (5) The pitted (62, 81),
perforated (84, and 61, section 16), and irregularly etched
(8) masses of calcareous rocks, developing the stratification (6-7, 89), joint-structure (88), and very often an
otherwise imperceptible concretionary structure (26,
section 14). The illustrations under this head might be
extended almost indefinitely, including, as they do, all
those cases where the quiet erosive action of air and
water has been influenced by the structure of the rocks.

Marine Erosion.—The waves and currents of the
ocean and inland seas and lakes develop the rounded
forms of pebbles (23) from angular fragments of rock ;
smooth and round the ledges where they can wash sand
or gravel over or against them ; and more truly etch the
rocks where the water acts alone, carefully discriminating

between the hard and soft parts (85–86). The large specimen from Deer Isle, Maine, is a remarkably fine ex‑ ample of the sculpturing produced when the waves beat directly against a cliff of yielding and variable rock. In such cases the results resemble those due to chemical erosion.

Fluvial Erosion. — Running water, in brooks and rivers, produces pebbles similar to those of the beach ; it smooths and rounds, or etches the ledges in the same way, and almost the only distinctive feature is the pot‑ holes, which are far more characteristic of the torrential portions of rivers than of the shores of lakes and seas. They are formed commonly in the solid ledge, but some‑ times in bowlders of suitable form (81) ; and of course only where the water has sufficient force to rotate stones which have collected in a hollow in the surface of the rock.

Glacial Erosion.—A glacier smooths and polishes (45) or striates (48–51) the rocks over which it moves and tends to reduce them to a plane surface ; its erosive action, unlike that of water, being relatively indiscrim‑ inating. In cases, however, where the rock presents un‑ usual contrasts in hardness, the harder portions will be left in slight relief. Thus in the strongly glaciated frag‑ ment of coarse vein granite (44) the resistant masses of quartz protrude slightly beyond the more yielding feld‑ spar. Where the protruding portions or hummocks of a glaciated surface are more pronounced, the northern or stoss side presents, normally, a long, smooth, and strong‑

ly glaciated slope, and the southern or lee side a short, abrupt, and craggy or relatively unglaciated slope. By attention to these details, it is often possible to determine, even in the case of detached specimens, in which way along a series of striae the ice moved. When developed on a larger scale, these protruding masses or ledges with an unsymmetrical north-south profile are known as *roche moutonées*, of which there are many fine examples among the ledges of the Boston Basin (**47**). The stones dragged along by the moving ice not only wear away and striate the solid ledges, but they are themselves similarly smoothed and scratched on their working faces (**42–43**) ; and the glaciated or ice-worn stones are easily distinguished from the water-worn stones or pebbles (**23**) by the parallel striae, and by the fact that they are not usually glaciated on all sides. If a stone is turned in the ice, so as to be worn on several sides, it is not rounded ; for glaciation develops a flat surface or facet, and such stones may be described as facetted (**46**). The large mass of slate from East Boston on the right side of the stairway, in the Vestibule, is a larger but very typical example of a glaciated bowlder ; and similar masses, five to ten feet in diameter, are common in excavations about Boston.

The pictures of the glacial pot-holes in Cohasset (**41**) might be duplicated at several other points in the Boston Basin. They are similar in their characteristics to ordinary river pot-holes, except that they usually occur in positions remote from streams, and where it is impossible that they should have been formed by ordinary brooks or rivers. They are, in fact, the product of gla-

cial mills (moulins) ; that is, they are formed where a
stream flowing over the surface of a glacier plunges
through a crevasse upon the solid ledges below.

The striated slab of sandstone and the bowlder of gneiss
from the Catskill Mountains (**82–83**, section 16) are of
interest especially because they were found near the
upper limit of glaciation in that region, about 3,000 feet
above the sea, and the bowlder has very clearly been
brought by the ice from the Adirondacks, nearly one
hundred miles to the north. The bowlder from the summit of Mt. Washington (**84**) is wholly unlike the rock in
place on this mountain (**85**), and thus proves that the great
ice-sheet covered the highest summit of New England.

Aerial Erosion.—The wind has considerable power
to wear away and transport unconsolidated sands ; and
when thus armed with moving sand it is also able, as in
the case of the artificial sand-blast, to wear away the
hardest and most resistant rocks. The results of the
natural sand-blast, as seen in the polished and striated
specimens of a hard quartzose rock from Nevada (**26**),
may simulate glaciation very closely. The smooth but
angular pebbles from Colorado (**28**) and Nantucket (**24–
25**) are believed to have been fashioned by blown sands,
although it is not improbable that the solvent action of
water has played some part here. The angular sculpturing is, as a rule, limited to the upper sides of the pebbles.

Organic Erosion.—The only clear illustrations to be
noted here are the rock-borings of certain bivalve shells
(**29–31**) and the spiny echinoderm or echinus (**27**).

Major Erosion Features.

Turning now to erosion-forms in their larger or topo-graphic aspect, the photographs in this section may be referred to as a mere suggestion of what might be done, if space permitted, with this class of illustrations. But even these show some of the more distinctive features of marine (**1**), fluvial (**2**), sub-aerial (**21-23**), and glacial (**42**), erosion, as explained on the labels. The relief map of the Mt. Blanc range (**61**) may be most conveniently noticed in this connection. In spite of its mantle of ice and snow, its contours show the normal results of subaerial erosion at a great elevation on massive rocks. In addition, it is an admirable and comprehensive illustration of local, as contrasted with general, glaciation.

Horizontal or very slightly undulating strata, especially if the upper beds are harder than those below, give rise by erosion to flat-topped ridges or table mountains. But if the strata be softer and of more uniform texture, erosion yields rounded hills, often very steep, and sometimes passing into pinnacles, as in the Bad Lands of the West. Broad, open folds, give, normally, synclinal hills and anticlinal valleys, when the erosion is well advanced. But in more strongly, closely folded rocks the ridges and valleys are determined chiefly by the outcrops of harder and softer strata, the symmetry of the reliefs often depending upon the dip of the strata. This principle of unequal hardness or durability also determines most of the topographic features in regions of metamorphic and crystalline rocks, in which the stratification is obscure or wanting.

The boldness of the topography, and the relation of
depth to width in valleys, depends largely upon the altitude above the sea ; but partly, also, upon the distribution
of rain-fall, the drainage, channels or valleys being narrowest and most sharply defined in arid regions traversed
by rivers deriving their waters from distant mountains.
That these are the conditions most favorable for the formation of cañons is proved by the fact that they are fully
realized in the great plateau country traversed by the
Colorado and its tributaries, a district which leads the
world in the magnitude and grandeur of its cañons. See
the large relief map in the window space. But deep
gorges and cañons will be formed wherever a considerable
altitude, by increasing the erosive power of the streams,
enables them to deepen their channels much more rapidly
than the general face of the country is lowered by rain
and frost. This is the secret of such cañons as the gorge
of the Columbia River, and probably of the fiords which
fret the northwest coasts of this continent and Europe.

The smaller plaster model (**22**) with the accompanying
section directs attention to the fact that erosion is not
limited to the surface of the earth, but, in the case of limestones especially, is very largely subterranean. The solvent action of underground waters, as explained on the
label, leads to the formation of caverns, natural bridges,
sink-holes and eventually to the lowering of the surface
over extensive areas.

The two wooden models (**41, 43**) which are the exact
complements of each other, one representing a ridge and
the other a valley cut out of horizontal strata, are intended
especially to illustrate certain relations of strata to the

topography to which attention must be given in the construction of geologic maps. These are : First, that the breadth of the outcrops of strata are inversely proportional to the inclination of the surface, the outcrop being narrowest where the slope is steepest, and *vice versa*. This is the rule for horizontal strata ; with highly inclined or vertical strata it is reversed. Second, that the outcrops of horizontal strata, being essentially like contour- or shore-lines, are deflected, in crossing the topographic forms, up the valleys and toward the lower ends of ridges.

From these general and somewhat isolated illustrations of topographic geology, we pass to the excellent series of topographic models or relief maps which are placed in the upper parts of the cases and on the walls of the window-spaces of this room. These were designed by Prof. W. M. Davis, and their special merit and interest is found in the fact that, occurring in series, they illustrate the development or life-history of topographic features, which is otherwise virtually impossible. They begin with section 1, and since there is but one model in each section, they may be conveniently referred to by the section numbers.

The first series, including the first five models (1–5), starts with a region newly elevated above the sea and devoid of relief features, and traces the normal development and succession of the topographic forms in the absence of pronounced contrasts in the geological structure. In the first model the main drainage channel of the distri , and a few tributary channels, are distinctly outlined, but the interstream surfaces are still broad and

288 STRUCTURAL GEOLOGY. Sections 1–5.

level. This is evidently a very youthful topography. In the second model (2) the development is much more advanced. The drainage lines of the first model are all easily recognized; but the tributaries, cutting steadily backward into the primitive plain, have been greatly extended and ramified; the valleys have been deepened, and broadened by the reduction of their lateral slopes; and the original interstream surfaces are reduced to isolated remnants or completely effaced. This evidently represents a period of topographic maturity, of maximum complexity and relief. This is followed, in the next model (3), by a subsidence of the land. The main valley and the lower portions of the tributary valleys are invaded by the sea and become an area of deposition; and while this arm of the sea is being silted up, marine, fluvial, and subaerial erosion are steadily wearing down the surrounding land. Thus by a two-fold process—filling up the valleys and degrading the hills—the topography is flattened and its features slowly effaced. The fourth model (4), represents a second elevation of the land; the sea has retired, and the streams have begun to re-excavate their well-nigh obliterated channels. This means, virtually, a renewal of the youthful conditions, the beginning of a new cycle of development, which proceeds (5) without interruption by a movement of subsidence until all the lower parts of the district are worn down nearly to the baselevel, and the general surface reduced to a peneplain (nearly a plain), the true topographic old age.

In the next series, including three models (6-8), the topography is, obviously, controlled in a much larger degree by the geologic structure. The region represented is one traversed by approximately parallel ranges of mountains, which have, like the parallel ranges of the Great Basin, been produced in the first instance by fault-ing, by the tilting of long and narrow blocks of the earth's crust, the precipitous faces of the ranges being the somewhat eroded fault scarps.   In this case the valleys and all the main topographic features, are clearly ante-cedent, and not the product of erosion.   The first model represents the valleys as partially filled with water, re-sembling the extensive Quaternary lakes of the Great Basin.   In these lakes silt washed from the mountains is deposited in horizontal layers to a great depth.   The cli-mate then becomes more arid, the lakes waste by evapor-ation and are finally reduced to a series of isolated and shallow pools, as shown in the second model (7), in which the buff tint represents the newly-deposited sediments. Most of these pools or playas, being without outlets, are necessarily saline or alkaline.   But as the small streams draining into them cut down and cut back their channels, the playas of the same valley are gradually drained, and the whole valley eventually drains directly into one basin or pool.   The principal streams tributary to such a basin will naturally rise in the passes or low gaps between the ranges that bound the valley ; and the head waters of the streams, cutting steadily backward, will sometimes reduce these barriers sufficiently so that the stream in the higher valley will be gradually reversed and become tributary to the lower valley.   Through the continued and repeated operation of this principle, all the valleys represented ultimately become united in one connected but intricate drainage system, as shown in the third model.   The

principle especially illustrated by this series is, evidently
the union of distinct, and even closed, drainage systems
or hydrographic basins in one continuous system, through
the backward erosion of the streams, the basins having
been made distinct in the first place by the deformation
of the earth's crust, by causes quite independent of
erosion.   A river system thus tends to become constantly
more extended, through the backward cutting of its
headwaters and the development of new tributaries or
branches, and also by capturing a part or the whole of
adjacent systems and thus enlarging its drainage area.
Streams of considerable fall and erosive power possess,
evidently, a marked advantage, in these respects, over
those of more sluggish habit.

The next series embraces but two models (9-10) and
these are intended especially to illustrate the topographic
contrast between a typical unglaciated area (9) and a
typical glaciated area (10) ; or, in other words, to show
the effect of glaciation upon a topography which has
been fully and normally developed by aqueous erosion.
In the first model the topographic development has
passed the period of maximum ruggedness, the original
interstream surfaces having been entirely obliterated ; but
the drainage over the entire area is perfect, free and
unobstructed.   The second model represents precisely
the same area after long-continued glaciation, during which the hills have lost still more of their rug-
ged character and the valleys have become clogged
with irregular deposits of drift, greatly obstructing and even diverting the drainage.   The obstruction
of the drainage is seen in the numerous lakes and lake-

lets, some of which have no outlets ; and a careful comparison with the first model shows not only that many of
the minor branches have been obliterated or greatly modified, but also that one of the largest tributaries has been
diverted near its junction with the main stream and
forced to add its waters to those of a smaller tributary,
finally reaching the main stream at a point considerably
below its original confluence. The diversion of the
streams will often cause them to flow over ledges or
ridges of solid rock, giving rise to waterfalls, which are
wanting in the unglaciated landscape. Eventually, however, the streams, by deepening their channels, will
drain all the standing water, and obliterate the cascades ;
and their altered courses, with remnants of drift upon
the hills, will alone remain to tell of this crisis in their
history.

We pass now to a series of six models (**11–16**) illustrating the influence of volcanic phenomena upon the
topography, the illustration embracing the entire life-
history, or development and decay, of two volcanoes.
The first model (**11**) represents a flat country with
slightly incised drainage lines having a general east to
west course. Volcanic activity commences near the
sources of the streams (**12**), a cinder cone (green color)
is formed and a stream of lava flows southward and occupies the valley of the southern stream, damming back
its waters and forming two lakes. In the third model
(**13**) these lakes have been drained by the erosive action
of the stream flowing from them along the edge of the
lava. But a more recent flood of lava, having a general southwesterly course, has invaded this stream at a

lower point, intercepting its main tributary and damming
it back until its waters discharge over the low water-
parting into the northern stream. Two small lakelets
are also formed at the confluence of the two lava streams
in the northern valley. Two flows of lava have also invaded the northern valley, slightly impounding its waters.
In the fourth model (**14**) this crater is represented as
extinct and holding a lakelet; and all but one of the
lakes formed by its flows have been drained. Meanwhile, however, a second cone has arisen in the eastern
part of the field, and from this new center of volcanic
activity lava has spread widely over the surrounding
country, obstructing the northern stream and obliterating the large lake formerly existing between the two
cones. In the fifth model (**15**) this second crater is also
extinct, both cinder cones are largely wasted by erosion,
and in the older one the central core or plug, the volcanic "neck" begins to appear. The lakes are all
drained; but the flows of solid, resistant lava are still almost intact, although the streams which border them are•
steadily gnawing at their edges. In the last model (**16**)
the cinder cones have wholly disappeared, the protruding necks alone remain to mark their sites, and the
lateral erosion of the streams has greatly narrowed the
sheets of lava, which, although formed in the valleys,
now rise abruptly above the surrounding country.

The next series embraces the two models on the left
side of the south window   These are intended especially
to show the normal relations of erosion to anticlinal and
synclinal folds, and may be regarded as especially typical of a large area of mountainous country in middle

Pennsylvania. The lower model is purely ideal, representing the strata as bent into a series of sharp parallel folds without having suffered erosion. The agreement between the geology and topography is perfect, each ridge corresponding to an anticline and each valley to a syncline. Now the rocks on the crest of the anticline have been stretched and broken, while those forming the trough of the syncline have been compressed and made more resistant; and since the anticline is also more elevated and more exposed to erosion, it follows that it will usually be worn down more rapidly than the syncline. And if the upper strata are underlain by softer rocks, as soon as the former are worn through on the anticline the erosion will proceed still more rapidly, the crests of the anticlines being very generally replaced by narrow valleys, as shown in the upper model, in which the valleys are about equally divided between the synclinal valleys due to the original folding of the strata and the anticlinal valleys due to subsequent erosion. The synclinal or structural valleys would, in many cases, be without outlets, but for the transverse notching of the ridges by the streams; and it will be observed that in several instances the anticline is now more deeply eroded than the adjacent syncline, foretelling a time when the synclinal valleys will have disappeared and the relations of topography to geology be completely reversed or transposed, — anticlinal valleys and synclinal ridges.

The next series includes the two models on the right side of the south window. It expresses still another relation of topography to geology. The lower model represents a great mountain uplift, the combined product of gigantic faults and flexures, and entirely unmodified by erosion. It shows what the subterranean agencies would have produced if the superficial agencies had, meanwhile, been at rest. The upper model shows the same uplift modified by erosion ; and in the river which cuts directly across the uplift or mountain range, we have evidence not only that the erosion was in progress during the uplifting, but also that the river is older than the uplift, and that the elevation of the mountain took place so slowly that the river was able to keep pace with it and preserve its course. The horizontal and sloping terraces afford a clue to the stratigraphy of the uplift.

The next pair of models, on the left side of the middle window, is intended especially to illustrate the capture by one stream of part of the drainage area of another stream. Two streams are represented, one flowing north and the other south, their basins or drainage areas being distinguished by the different colors. The headwaters of the northward-flowing stream, as shown in the lower model, have much the greatest fall and erosive power, and in their backward gnawing they eventually capture, as shown in the upper model, the first two branches of the southward-flowing stream. The sluggish character of the latter is seen in the fact that, after being deprived of a good part of its headwaters, it is no longer able to sweep away the detritus brought down by its lateral branches, which clogs the valley and forms shallow ponds of standing water.

The pair of models on the right side of the middle window is another illustration of the same general character and purpose ; one tributary of a main stream stealing water from another tributary. The lower model should be observed first, and then compared with the upper model, noticing the increased extent of the yellow tint and the diminished extent of the gray. In the next pair of models, on the left side of the north window, the same general story is told again. The left hand model shows a large stream and several brooks flowing into a lake, the large stream having commenced to form a delta. In the other model one of the brooks has succeeded in capturing all the upper part of the large stream, which below the scene of the piracy has become relatively insignificant, and delta-deposits have become more general along the coast.

The three models in the passage way to Room A illustrate the river terraces so characteristic of the valleys running south from the retreating front of the great ice-sheet. The first model of the series shows the valley before the ice age. The enormous floods of water resulting from the melting of the ice washed down vast amounts of sand and gravel which were deposited in the valleys, filling them up in some cases to a depth of several hundred feet, as shown in the second model.

These accumulations dammed the mouths and impounded the waters of tributary valleys, which were not swept by the glacial torrents. After the ice-sheet had passed away and the streams had regained their normal dimensions and were no longer overloaded with detritus, they began to clear out their channels; and the progress which they have made in this work is recorded in the successive terraces which rise, step-like, on either side of the valley to the hight of the glacial flood-plain, as shown in the third model. The terraces below the highest are simply remnants of later and still later flood-plains down to the lowest, which is the modern flood-plain of the river.

INDEX.

(297)

www.ingramcontent.com/pod-product-compliance
Lightning Source LLC
Chambersburg PA
CBHW021504210326

41599CB00012B/1135